OXFORD CHEMISTRY PRIMERS

58

PHYSICAL

Statistical Thermodynamics

Andrew Maczek

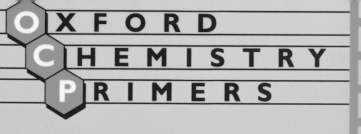

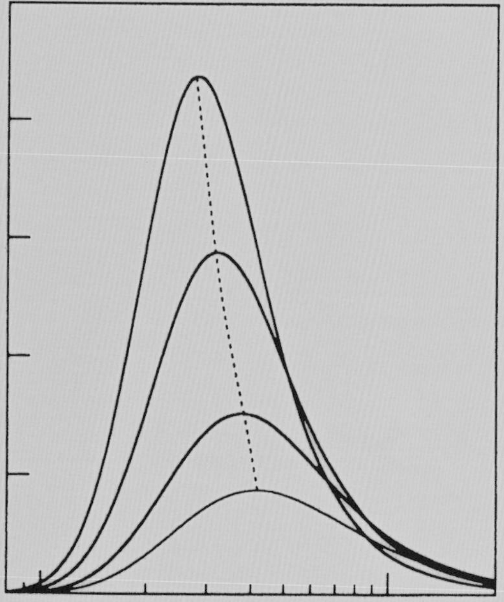

OXFORD SCIENCE PUBLICATIONS

OXFORD CHEMISTRY PRIMERS

1 S. E. Thomas *Organic synthesis: The roles of boron and silicon*
2 D. T. Davies *Aromatic heterocyclic chemistry*
3 P. R. Jenkins *Organometallic reagents in synthesis*
4 M. Sainsbury *Aromatic chemistry*
5 L. M. Harwood *Polar rearrangements*
6 T. J. Donohoe *Oxidation and reduction in organic synthesis*
7 J. H. Jones *Amino acid and peptide synthesis* **Second edition**
8 C. J. Moody and G. H. Whitham *Reactive intermediates*
9 G. M. Hornby and J. M. Peach *Foundations of organic chemistry*
10 R. Henderson *The mechanisms of reactions at transition metal sites*
11 H. M. Cartwright *Applications of artificial intelligence in chemistry*
12 M. Bochmann *Organometallics 1: Complexes with transition metal–carbon s-bonds*
13 M. Bochmann *Organometallics 2: Complexes with transition metal–carbon p-bonds*
14 C. E. Housecroft *Cluster molecules of the p-block elements*
15 M. J. Winter *Chemical bonding*
17 R. S. Ward *Bifunctional compounds*
18 S. K. Scott *Oscillations, waves, and chaos in chemical kinetics*
19 T. P. Softley *Atomic spectra*
20 J. Mann *Chemical aspects of biosynthesis*
21 B. G. Cox *Modern liquid phase kinetics*
22 A. Harrison *Fractals in chemistry*
23 M. T. Weller *Inorganic materials chemistry*
24 R. P. Wayne *Chemical instrumentation*
25 D. E. Fenton *Biocoordination chemistry*
26 W. G. Richards and P. R. Scott *Energy levels in atoms and molecules*
27 M. J. Winter *d-Block chemistry*
28 D. M. P. Mingos *Essentials of inorganic chemistry 1*
29 G. H. Grant and W. G. Richards *Computational chemistry*
30 S. A. Lee and G. E. Robinson *Process development: Fine chemicals from grams to kilograms*
31 C. L. Willis and M. R. Wills *Organic synthesis*
32 P. J. Hore *Nuclear magnetic resonance*
33 G. H. Whitham *Organosulfur chemistry*
34 A. C. Fisher *Electrode dynamics*
35 G. D. Meakins *Functional groups: Characteristics and interconversions*
36 A. J. Kirby *Stereoelectronic effects*
37 P. A. Cox *Introduction to quantum theory and atomic structure*
38 P. D. Bailey and K. M. Morgan *Organonitrogen chemistry*
39 C. E. Wayne and R. P. Wayne *Photochemistry*
40 C. P. Lawrence, A. Rodger, and R. G. Compton *Foundations of physical chemistry*
41 R. G. Compton and G. H. W. Sanders *Electrode potentials*
42 P. B. Whalley *Two-phase flow and heat transfer*
43 L. M. Harwood and T. D. W. Claridge *Introduction to organic spectroscopy*
44 C. E. Housecroft *Metal–metal bonded carbonyl dimers and clusters*
45 H. Maskill *Mechanisms of organic reactions*
46 P. C. Wilkins and R. G. Wilkins *Inorganic chemistry in biology*
47 J. H. Jones *Core carbonyl chemistry*
48 N. J. B. Green *Quantum mechanics 1: Foundations*
49 I. S. Metcalfe *Chemical reaction engineering: A first course*
50 R. H. S. Winterton *Heat transfer*
51 N. C. Norman *Periodicity and the s- and p-block elements*
52 R. W. Cattrall *Chemical sensors*
53 M. Bowker *The Basis and applications of heterogeneous catalysis*
54 M. C. Grossel *Alicyclic chemistry*
55 J. M. Brown *Molecular spectroscopy*
56 G. J. Price *Thermodynamics of chemical processes*
57 A. G. Howard *Aquatic environmental chemistry*
58 A. O. S. Maczek *Statistical thermodynamics*
59 G. A. Attard and C. J. Barnes *Surfaces*
60 W. Clegg *Crystal structure determination*
61 M. Brouard *Reaction dynamics*
62 A. K. Brisdon *Inorganic spectroscopic methods*
63 G. Proctor *Stereoselectivity in organic synthesis*
64 C. M. A. Brett and A. M. O. Brett *Electroanalysis*
65 N. J. B. Green *Quantum mechanics 2: The toolkit*
66 D. M. P. Mingos *Essentials of inorganic chemistry 2*
67 I. Fleming *Pericyclic reactions*
68 N. S. Lawrence, J. D. Wadhawan & R. G. Compton *Foundations of physical chemistry: Worked examples*
69 J. R. Chipperfield *Non-aqueous solvents*
70 T. J. Mason *Sonochemistry*
71 J. McCleverty *Chemistry of the first-row transition metals*
72 E. C. Constable *Coordination chemistry of macrocyclic compounds*
73 C. E. Housecroft *The heavier d-block metals: Aspects of inorganic and coordination chemistry*
74 P. D. Beer, P. A. Gale and D. K. Smith *Supramolecular chemistry*
76 N. Kaltsoyannis and P. Scott *The f elements*
77 D. S. Sivia and S. G. Rawlings *Foundations of science mathematics*
78 S. Duckett and B. Gilbert *Foundations of spectroscopy*
79 J. H. Atherton and K. J. Carpenter *Process development: Physicochemical concepts*
80 R. de Levie *Acid–base equilibria and titrations*
81 H. Maskill *Structure and reactivity in organic chemistry*
82 D. S. Sivia and S. G. Rawlings *Foundations of science mathematics: Worked problems*
83 J. Iggo *NMR Spectroscopy in inorganic chemistry*
84 P. Biggs *Computers in chemistry*
85 D. J. Walton and P. Lorimer *Polymers*
86 R. J. Davey and J. Garside *From molecules to crystallizers*
87 G. M. Hornby and J. M. Peach *Foundations of organic chemistry: Worked examples*
88 M. J. T. Robinson *Organic stereochemistry*
89 H. R. N. Jones *Radiation heat transfer*
90 J. Saunders *Top drugs—top synthetic routes*
91 M. J. Perkins *Radical chemistry: The fundamentals*
92 P. J. Hore, J. A. Jones, and S. Wimperis *NMR: The toolkit*
93 G. A. D. Ritchie and D. S. Sivia *Foundations of physics for chemists*
94 M. J. Winter and J. Andrew *Foundations of inorganic chemistry*
95 J. Robertson *Protecting group chemistry*
96 R. Whyman *Applied organometallic chemistry and catalysis*
97 J. S. Ogden *Introduction to molecular symmetry*
98 C. M. Dobson, A. J. Pratt, and J. A. Gerrard *Bioorganic chemistry*
99 B. G. Davis and A. J. Fairbanks *Carbohydrate Chemistry*

Statistical Thermodynamics

Andrew Maczek

Senior Lecturer, Department of Chemistry, University of Sheffield

OXFORD
UNIVERSITY PRESS

OXFORD
UNIVERSITY PRESS
Great Clarendon Street, Oxford OX2 6DP
Oxford University Press is a department of the University of Oxford.
It furthers the University's objective of excellence in research, scholarship,
and education by publishing worldwide in

Oxford New York

Auckland Cape Town Dar es Salaam Hong Kong Karachi
Kuala Lumpur Madrid Melbourne Mexico City Nairobi
New Delhi Shanghai Taipei Toronto

With offices in

Argentina Austria Brazil Chile Czech Republic France Greece
Guatemala Hungary Italy Japan Poland Portugal Singapore
South Korea Switzerland Thailand Turkey Ukraine Vietnam

Published in the United States
by Oxford University Press Inc., New York

Reprinted 2002, 2003, 2004, 2005, 2006, 2007

A catalogue record for this book is available from the British Library

Library of Congress Cataloging in Publication Data
(Data available)

ISBN 978-0-19-855911-5

7 9 10 8

Typeset by the Author

Printed in Great Britain by Antony Rowe Ltd,
Chippenham, Wiltshire

Series Editor's Foreword

Oxford Chemistry Primers are designed to provide clear and concise introductions to a wide range of topics that may be encountered by chemistry students as they progress from the freshman stage through to graduation. The Physical Chemistry series aims to contain books easily recognised as relating to established fundamental core material that all chemists need to know, as well as books reflecting new directions and research trends in the subject, thereby anticipating (and perhaps encouraging) the evolution of modern undergraduate courses.

In this Physical Chemistry Primer Andrew Maczek provides a simple and clearly explained introduction to the topic of *Statistical Thermodynamics* together with some of its applications. This is an area of central importance in Physical Chemistry and forms an essential part of all undergraduate courses in this area. Accordingly this Primer will be of interest to all students of chemistry and their mentors.

Richard G. Compton
Physical and Theoretical Chemistry Laboratory
University of Oxford

Preface

Statistical Thermodynamics is, in its fullest sense, a *complete* subject. It spans the extremes from Classical Thermodynamics, which firmly avoids any reference to the underlying microscopic structure of matter, through Spectroscopy, whose interpretation requires a willingness to visualise matter in terms of atoms and molecules, to Quantum Mechanics which illuminates the understanding of spectral behaviour. And it leads to such productive ends! It provides an ability to calculate equilibrium constants from spetroscopic data using classical thermodynamics, and also (beyond the scope of this text) to the possibility of predicting the rate at which equilibrium is attained using Activated Complex Theory. It is, above all, a *unifying* topic in Physical Chemistry.

Some, coming to Statistical Thermodynamics for the first time, find its rigour, coupled to its simplicity, a little daunting. This book attempts to dispel such worries and to present the topic in a logical and manageable manner. From the outset, in Chapter 1, where the concept of the *most probable configuration* is introduced, to the conclusion, in Chapter 15, where the *position of equilibrium* in real systems is calculated, the reader is encouraged to approach the subject through a series of short, self-contained, sections. It is hoped that this approach will make the subject more transparent and easier to understand.

To those who have encouraged me to write this book I am truly grateful. I owe most, perhaps all, to those of my students who, over the years, have reassured me by asking the right questions and giving the right answers. But also, I acknowledge two of my mentors, Peter Gray and Courtenay Phillips, who were the first to show me that it could all be such fun.

Department of Chemistry, University of Sheffield A. O. S. M.
October 1997

Contents

1 The Boltzmann law

1.1 Introduction

The objective of this book to introduce the reader to the methods that link all the predictive potential of classical thermodynamics to an existing understanding of the quantum nature of matter. Classical thermodynamics, based as it is on working fluids which are featureless and lack internal structure, enables us to deduce hitherto unmeasured quantities for specified substances and to predict the position of the equilibrium they will eventually reach if mixed under specified conditions.

Thermodynamics is based on three laws of experience. Once these laws are accepted, the edifice it builds is totally self-consistent. If a given measurement has the value X under one set of conditions and the value Y under another, then a number of inescapable consequences follow. In particular, if we can gather together comprehensive lists of free energy data, as is done in thermodynamic tables, we can find ourselves in a position to be able to predict, in quantitative detail, much of what lies in the field we call *chemical change*.

A law of experience is one that is based on observation. It persists only as long as no evidence is found to refute it.

What is lacking, however, is any clear view of why free energies take the values they do. It would be immensely satisfying to be able to predict these values directly from known molecular properties, without having to repeat specific calorimetric measurements of the kind that have given validity to the laws of thermodynamics. It is this predictive power, the ability to link function to structure, that statistical thermodynamics addresses and, to a large extent, succeeds in achieving.

1.2 The Boltzmann factor

At the heart of much of physical chemistry lies the exponential factor

$$\text{Physical quantity} \propto \exp\left[-\frac{f_E(E)}{\text{const}\times f_T(T)}\right] \tag{1.1}$$

where f_E and f_T are functions of energy and temperature only and the constant, with units *energy per unit of temperature*, is required to make the exponent dimensionless. This expression can be linked directly to the Boltzmann factor, $e^{-\varepsilon/kT}$ and, in turn, to the Boltzmann law (eqn 1.2) for the population of quantised energy states, as shown in Fig. 1.1.

$$\frac{n_i}{n_j} = e^{-(\varepsilon_i-\varepsilon_j)/kT} \tag{1.2}$$

THE BOLTZMANN LAW

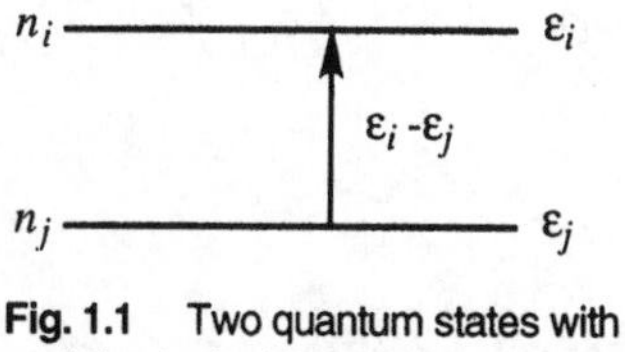

Fig. 1.1 Two quantum states with n_j particles in the lower state with energy ε_j and n_i particles in the upper state with energy ε_i.

In SI units, the temperature, T, is in kelvins and energies are expressed in joules. The constant of proportionality, the Boltzmann constant, k, takes the value 1.38×10^{-23} J K^{-1}.

It is the objective of this chapter briefly to outline the basis of the Boltzmann law before applying it in subsequent chapters. At some stage in a scientific education, it is necessary to confront and to appreciate the origins of this law, but its application to the problems we meet depends less on how it is derived and much more on how it can be used. For those who have already met the derivation, it is perfectly reasonable to turn directly to Chapter 2 and seek such applications.

1.3 The average basis of the behaviour of matter

The first, and vital, step in linking the individual properties of molecules and the thermodynamic behaviour of bulk matter is to realise that thermodynamic properties are concerned with **average behaviour**. Properties such as temperature and pressure (and others), which are characteristic of large assemblies of molecules, are average properties. At any instant, the actual pressure exerted on a given wall may fluctuate a little from the average as slightly more or slightly fewer molecules happen to strike that wall. The actual temperature in one small part of the total volume may be a little higher than the average because of a fluctuation leading to more of the high speed molecules (which carry more energy and are "hotter") congregating there. But such fluctuations turn out to be negligibly small compared with the established average, simply because the number of particles contributing to the average is so vast.

In this sense, the Boltzmann law (eqn 1.2) is also a law that deals with averages. To be strictly correct, the occupation numbers of the two states should be written as averages $\langle n_j \rangle$ and $\langle n_i \rangle$, as in eqn 1.3, since these numbers too are subject to infinitesimal fluctuations

$$\frac{\langle n_i \rangle}{\langle n_j \rangle} = e^{-(\varepsilon_i - \varepsilon_j)/kT} \tag{1.3}$$

However, since the instantaneous values of the occupation numbers are never very different from the averages, we shall dispense with stressing their average nature in everything that follows.

1.4 Plan of attack

In order to gain an understanding of how the Boltzmann law can be reached, we need to use certain of the laws of probability, to establish some assumptions and to understand some of the language that is in accepted use. The list below can serve as a route-map of the progress we hope to make, and each item is discussed in turn in the sections that follow.

- distinct, independent particles
- configurations: sharing out the energy
- statistical weights
- equal probability of configurations
- conservation of number and energy
- the predominant configuration
- maximisation subject to constraints

1.5 Distinct, independent particles

We start by assuming that we have an assembly of particles at **constant temperature.** These particles are **distinct** in the sense that we can always tell which is which They carry labels (a, b, c, ..., etc). They are also **independent**. We also assume that they interact with each other minimally, just enough to interchange energy at collision. We say that they are **weakly coupled**, and this allows us to write the total energy E, at any instant as the sum of the individual energies of these labelled particles, ignoring any mutual potential energies between them

The traditional way of ensuring a constant temperature is to immerse our assembly in a heat bath that is large enough to ensure that there is no overall fluctuation in temperature.

$$E = \varepsilon^a + \varepsilon^b + \varepsilon^c + \varepsilon^d + \ldots = \sum_i \varepsilon^i \qquad (1.4)$$

1.6 Configurations: sharing out the energy

Individual particles have access to a manifold of **energy states**, with energies ε_0, ε_1, ε_2, ε_3, ..., etc. From now on we shall adopt the convention that the state with lowest energy is the zero of energy, i.e. $\varepsilon_0 = 0$. This may mean that we have to add some constant value to thermodynamic energy functions such as U, H, A, ..., etc, which are calculated by summing over all possibilities, but this is only a minor inconvenience.

Conventionally $\varepsilon_0 = 0$

Thermodynamic energy functions:
- U, the internal energy
- H, the enthalpy
- A, the Helmholtz free energy
- G, the Gibbs free energy

Over a period of time, particles interchange energies (always subject to a constant total energy) so that, at any instant, the distribution of particles among energy states may involve, say, n_0 with energy ε_0, n_1 with energy ε_1, n_2 with energy ε_2, and so on. We call this instantaneous distribution the **configuration** of the system. At the next moment the distribution will be different, giving a different configuration with the same total energy. These configurations identify the way in which the system can share out its energy among the available energy states. For a large number of particles, many different configurations are possible.

1.7 Statistical weights

A given configuration can be reached in a number of different ways. We call the number of ways, Ω, the **statistical weight** of that configuration. It represents the **probability** that this configuration can be reached, from among all other configurations, by totally random means.

A useful analogy here is to make reference to coin-tossing. If we were to toss a hundred coins successively a huge number of times, then clearly the configuration "all heads" (or "all tails") would occur significantly less often than the configuration "50 heads, 50 tails". The importance of the analogy, however, is that we know how to calculate the probability (statistical weight) of any given configuration, "72 heads, 28 tails", say. For N particles arriving at a configuration in which there are n_0 particles with energy ε_0, n_1 with energy ε_1, n_2 with energy ε_2, and so on, the statistical weight is

$$\Omega = \frac{N!}{n_0!n_1!n_2!n_3! \ldots.} \qquad (1.5)$$

The factorial function:
$x! = x(x-1)(x-2)(x-3)(x-4)\ldots3.2.1$
By definition, $0! = 1$.

1.8 Equal probability of configurations

Principle of equal *a priori* probabilities.

We assume that there is no bias towards any configuration a system can adopt, that it is free to roam unhindered through all possible configurations. In the course of time (perhaps infinite time) each configuration will be visited exactly in proportion to its statistical weight. We call this the **principle of equal *a priori* probabilities**. We shall need to use it.

1.9 Conservation of number and energy

These are rather obvious constraints that we place on our system, but ones we shall need to invoke shortly. In symbolic terms

$$N = \sum_i n_i$$
$$E = \sum_i n_i \varepsilon_i \qquad (1.6)$$

1.10 The predominant configuration

The set of occupation numbers, n_0 with energy ε_0, n_1 with energy ε_1, n_2 with energy ε_2, and so on, describes just one configuration. How likely this configuration is depends on its statistical weight, and we know how to calculate this using eqn 1.5.

What we need to do now is to find the most probable configuration, the one with the largest statistical weight. We do this in the next section. But before doing so, we ask the question, how likely is the most probable configuration to dominate the assembly and impose on it its own characteristics? This question is a reasonable one, for we already believe, from (vicarious) experience that in tossing a hundred coins successively very many times, the 50–50 configuration will stand out well above the rest.

What would happen with a mole of coins? The answer is as simple and as stark as it is ultimately helpful. For a number of particles of the order of 10^{23}, the distribution of configurations peaks so sharply about the most probable configuration that, effectively, no other configuration gets so much as a look-in. The proof of this contention can be appreciated most readily by the lengthy process of determining the shape of the probability distribution of different configurations. This turns out to be a curve with so narrow and sharp a peak that it gives to the most probable configuration a statistical weight that truly towers over all other possible configurations.

So overwhelming is this dominance once we reach numbers of particles (coins) of the order of Avogadro's number, that it is possible to be entirely confident that, out of the truly vast total number of possible configurations accessible to a very large assembly of particles, an overwhelming majority arise from configurations very close to the predominant configuration. All these configurations are extremely similar to one another, so the behaviour and properties of the assembly are effectively indistinguishable from the behaviour and properties of the predominant configuration.

The proof of this contention adds little to its force, so we shall adopt it as a further axiom. But just how true it is can be seen if one compares the statistical weight of the most probable configuration, Ω_{max}, for one mole of particles with the statistical weight, Ω, of another configuration only very slightly different (1 part in 10^{10} on average). The probability declines by a factor of over 10^{400} – a factor that the deceptive power of exponent notation can easily deceive us into thinking we can grasp.

The axiom of the predominant configuration.

For 10^{23} particles, an average change of configuration of only 1 part in 10^{10} reduces the probability by

$$\frac{\Omega_{max}}{\Omega} \approx 10^{434}$$

a massive collapse in configuration probability.

Few configurations other than those indistinguishable from the most probable one contribute anything at all to the system. It is safe to assume that the properties of the most probable configuration are the only ones we need to consider. We discover these properties in the next section.

1.11 Maximisation subject to constraints

The rest of the task is comparatively simple. To discover the characteristics of the predominant configuration we need to find the maximum in the distribution of configurations (eqn 1.5) subject to the constraints imposed by eqn 1.6. This is an exercise in the calculus of conditioned maxima, the details of which need not concern us. They are amply set out in many standard texts, and lead to the conclusion that the predominant configuration among N particles has energy states that are populated according to the expression

$$\frac{n_i}{N} = e^{\alpha - \beta\varepsilon_i} \tag{1.7}$$

where α and β are constants under the conditions of constant temperature we chose in Section 1.5. We shall evaluate them shortly.

This last equation, only lightly disguised, is the Boltzmann law we set out to find. It provides the ultimate explanation of the pre-eminently important place that the exponential function plays in the whole of chemistry!

The constant α

This is easy to evaluate. If we adopt the convention we chose in Section 1.6, we can identify e^{α} with the fraction of particles in the ground state and write

The lowest state has energy $\varepsilon_0 = 0$ and occupation number n_0.

$$\frac{n_0}{N} = e^{\alpha} \tag{1.8}$$

which identifies the constant α and, combined with eqn 1.7, enables us to write

By the properties of exponents, $e^{\alpha - \beta\varepsilon_i} = e^{\alpha}.e^{-\beta\varepsilon_i}$.

$$\frac{n_i}{N} = e^{\alpha}e^{-\beta\varepsilon_i} = \frac{n_0}{N}e^{-\beta\varepsilon_i}$$
$$\therefore \quad \frac{n_i}{n_0} = e^{-\beta\varepsilon_i} \tag{1.9}$$

a temperature-dependent ratio since the occupation number of the lowest state, as well as that of the higher states, varies with temperature.

$\beta = \frac{1}{kT}$, see proof in Chapter 3

The constant β

It is less simple at this stage to determine the value of β. So, instead, we shall pre-empt the result (proved in Chapter 3) and simply state that

$$\beta = \frac{1}{kT} \tag{1.10}$$

where k is the Boltzmann constant, as before.

1.12 Conclusions

By invoking a series of reasonable assumptions, notably,

- distinguishable, independent particles
- equal probability of random configurations
- conservation of number and energy at constant temperature
- the predominance of the most probable configuration

we have been able to identify the occupation numbers of energy states and their temperature dependence in the predominant, and effectively singular, equilibrium configuration in a system of N particles at temperature T. This result is embodied in the Boltzmann law, from which we started, which tells us that for two states of different energy the ratio of the occupation numbers is given by eqn 1.2

$$\frac{n_i}{n_j} = e^{-(\varepsilon_i - \varepsilon_j)/kT} \tag{1.2}$$

This is the equation on which the rest of this text pivots.

2 Sum over states: the molecular partition function

2.1 Introduction

Starting with the statement of the Boltzmann law (eqn 1.2) which formed the cornerstone of our endeavours in Chapter 1

$$\frac{n_i}{n_j} = e^{-(\varepsilon_i - \varepsilon_j)/kT} \tag{1.2}$$

we now go on to investigate its properties and to derive from it another quantity, the **molecular partition function**, q, which will make it much simpler to apply the Boltzmann law to a variety of different situations. We start by establishing the meaning of the term *partition function*.

2.2 Occupation numbers, n_i, of molecular energy states

We start by rewriting eqn 1.2 with the lower energy state, (n_j, ε_j) as the state of lowest energy (n_0, $\varepsilon_0 = 0$) and with β replacing $1/kT$

$$n_i = n_0 e^{-\beta\varepsilon_i} \tag{2.1}$$

From equation 2.1, any state population (n_i) is known if ε_i, T, and n_0 are known.

Next we eliminate n_0, since this, in general, will not be known. If the total number of particles is N, then

$$N = n_0 + n_1 + n_2 + n_3 + \ldots = \sum_{\textit{all states}} n_i \tag{2.2}$$

and, using eqn 2.1

$$N = n_0 + n_0 e^{-\beta\varepsilon_1} + n_0 e^{-\beta\varepsilon_2} + n_0 e^{-\beta\varepsilon_3} + \ldots = n_0 \sum_{\textit{all states}} e^{-\beta\varepsilon_i} \tag{2.3}$$

where the summation extends over values of the exponential for all possible quantum states of the particles.

Rearranging, we get, for n_0

$$n_0 = \frac{N}{\sum_{\textit{all states}} e^{-\beta\varepsilon_i}} \tag{2.4}$$

From this we obtain an expression for the occupation number, n_i

$$n_i = \frac{Ne^{-\beta\varepsilon_i}}{\sum_{\text{all states}} e^{-\beta\varepsilon_i}} \tag{2.5}$$

or, writing

$$\sum_{\text{all states}} e^{-\beta\varepsilon_i} = q \tag{2.6}$$

we get

$$n_i = \frac{Ne^{-\beta\varepsilon_i}}{q} \tag{2.7}$$

which provides a compact expression for the occupation number, n_i,and defines the quantity q, called the **molecular partition function.**

2.3 The molecular partition function, *q*

The German name *Zustandsumme* is equally apt and means *sum-over-states.*

The molecular partition function, q, is aptly named because it determines how particles distribute (or partition) themselves over **accessible** quantum states. Thus we have for q

$$q = 1 + e^{-\beta\varepsilon_1} + e^{-\beta\varepsilon_2} + e^{-\beta\varepsilon_3} + \ldots \qquad (\varepsilon_0 = 0,\ \beta = \frac{1}{kT})$$

an infinite series that converges more rapidly the larger both the energy spacing between quantum states (ε_i) and the value of β is. Consequently, convergence is enhanced by lower temperatures since $\beta = 1/kT$.

In evaluating the partition function in practical cases, we can terminate the series as soon as $\beta\varepsilon \gg 0$ so that $e^{-\beta\varepsilon} \to 0$.

$q \to 1$ if $\varepsilon_1 \gg kT$.

If the first energy gap (ε_0 to ε_1) is large, and here (as in the rest of this text) *large* denotes that the energy in question, ε_1, is very much greater than the thermal energy, kT, then q tends to its lowest value, which is unity.

For successive energy gaps $\Delta\varepsilon$, $q \gg 1$ if $\Delta\varepsilon < kT$.

If, however, there are many quantum states with energies of the order of or less than the thermal energy, there will be many terms in the partition function with values of the order of unity and q can become very large.

q reflects the accessibility of quantum states.

So the magnitude of the partition function shows how easily particles spread over the available quantum states and thus reflects the **accessibility** of the quantum energy states of the particles involved.

Note that the partition function, although defined by eqn 2.5 in terms of an N particle assembly, has a value that is entirely independent of the actual value of N. We need N to be quite large in order for the occupation numbers of successive states to reflect truly the *equilibrium average* values that are of interest. So q is the partition function per particle, which is why we call it the molecular partition function.

2.4 Energy states and energy levels

The expression for the partition function given by eqn 2.6 can be modified, when it is convenient to do so, by taking into account that quantum states can be **degenerate**, with a number (g) of states all sharing the same energy. States with the same energy comprise an **energy level**, and we use the symbol g_i to denote the degeneracy of the i th level. Using this new terminology, for the partition function we can write

$$q = \sum_{\text{all levels}} g_j \, e^{-\beta \varepsilon_j} \tag{2.8}$$

an entirely equivalent way of formulating the partition function, with the running variable j serving to point to a distinction between *states* and *levels*.

2.5 The partition function explored

The total number of particles in our assembly is N or, expressed intensively, L per mole.

$$L = \sum_{\text{all states}} n_i = \sum_{\text{all levels}} g_j n_j = n_0 q \qquad \therefore q = \frac{L}{n_0}$$

so the partition function is a measure of the extent to which particles are able to escape from the ground state. The partition function q is a pure number which can range from a minimum value of 1 at 0 K (when $n_0 = L$ and only one state, the ground state, is accessible) to an indefinitely large number as the temperature increases (because fewer and fewer particles are left in the ground state and an indefinitely large number of states becomes accessible to the system).

The partition function for atoms which possess only translational energy can be of the order of 10^{30} at room temperature in a vessel of laboratory dimensions.

We can characterise the closeness of spacing in the energy manifold by referring to the **density of states** function, $D(\varepsilon)$, which represents the number of energy states in unit energy interval. If the density of states is high, particles will find it easy to leave the ground state and q will rise rapidly as the temperature rises. Conversely, a low density of states will lead to a small value of q ($\rightarrow 1$). The former is typical of translational motion in a gas, the latter of the vibrations of light diatomic molecules at room temperature.

Density of states: the number of energy states per unit energy interval.

However, if q/L, the number of accessible states per particle, is small, then few particles venture out of the ground state. If q/L is large, then there are many accessible states and molecules are well spread over the energy states of the system. We shall see later that the condition $q/L \gg 1$ is an important criterion for the valid application of the Boltzmann law in gaseous systems.

2.6 Conclusions

In this chapter we have defined the molecular partition function and have explored some of its important properties. In subsequent chapters we shall see that the molecular partition function has a vital role to play in thermodynamic calculations on a macroscopic assembly of particles, using spectroscopic measurements to characterise the molecular parameters of the particles that make up the assembly.

3 Applications of the molecular partition function

3.1 Introduction

In this chapter, we show how it is possible to use the molecular partition function to calculate a value for the **internal energy**, U, and the **entropy**, S, of an assembly of molecules.

3.2 The molecular energy, E

The total energy, E, of an assembly of distinct and independent particles is

$$E = n_1\varepsilon_1 + n_2\varepsilon_2 + n_3\varepsilon_3 + \ldots = \sum_{states} n_i\varepsilon_i \tag{1.6}$$

and, substituting values of the n_i from eqn 2.5, we have

$$E = \frac{\varepsilon_1 N e^{-\beta\varepsilon_1}}{\sum e^{-\beta\varepsilon_i}} + \frac{\varepsilon_2 N e^{-\beta\varepsilon_2}}{\sum e^{-\beta\varepsilon_i}} + \ldots = \frac{N\sum \varepsilon_i e^{-\beta\varepsilon_i}}{\sum e^{-\beta\varepsilon_i}} \tag{3.1}$$

The differential with respect to β of a single exponential term in the summation in eqn 3.1 is

$$\frac{d}{d\beta}(e^{-\beta\varepsilon_i}) = -\varepsilon_i e^{-\beta\varepsilon_i}$$

Introducing summations, this can be rearranged to give

$$\frac{d}{d\beta}\left(\sum e^{-\beta\varepsilon_i}\right) = -\sum \varepsilon_i e^{-\beta\varepsilon_i}$$

Thus the summation in the numerator of eqn 3.1 can be expressed in terms of the derivative of the partition function with respect to β.

In order to express the energy entirely in terms of the molecular partition function, we note the differential shown alongside. From this, introducing the molecular partition function, q, from eqn 2.6, it follows that

$$E = -\frac{N}{q}\frac{dq}{d\beta} = -N\left(\frac{d\ln q}{d\beta}\right) \tag{3.2}$$

3.3 The internal energy, U

Equation 3.2 needs two modifications before it can be used to yield the internal energy, U. The first concerns our choice $\varepsilon_0 = 0$ for the zero of energy. Equation 3.2 requires that $E = 0$ when $T = 0$, but the conventional internal energy, U, does not have this constraint. So, to obtain U from E, we must add the quantity $U(0)$, the internal energy at $T = 0$.

$$U = U(0) + E \tag{3.3}$$

so that

$$U = U(0) - N\left(\frac{d\ln q}{d\beta}\right) \tag{3.4}$$

The second modification recognises that q depends on quantities other than the temperature, since the energies of quantum states may vary with the volume occupied by the assembly, and V may vary with T. We can avoid this complication by specifying that the differentiation in eqn 3.4 must be carried out at **constant volume**.

$$U = U(0) - N\left(\frac{\partial \ln q}{\partial \beta}\right)_V = U(0) + NkT^2\left(\frac{\partial \ln q}{\partial T}\right)_V \tag{3.5}$$

Equation 3.5 shows that we need only know the molecular partition function and its temperature dependence in order to calculate the internal energy. In Chapter 7 we will show that all thermodynamic functions can be determined from a knowledge of the molecular partition function.

3.4 The relationship of β to temperature

The partition function for translational energy is derived in Chapter 8, but we shall use that result here to establish the relationship between β and T. In eqn 8.8, it is shown that the only temperature-dependent term in the partition function for kinetic (translational) energy in an N-particle assembly is

$$\ln q = -\frac{3}{2}\ln\beta + \text{ other terms independent of } \beta$$

which, in conjunction with eqn 3.5, leads to

$$U = U(0) + \frac{3}{2}N\left(\frac{\partial \ln \beta}{\partial \beta}\right)_V = U(0) + \frac{3N}{2\beta} \tag{3.6}$$

If we compare this with the internal energy of N atoms of a perfect gas, then the link between β and the absolute temperature is quite clear

$U = U(0) + \frac{3}{2}NkT$

$$\beta = \frac{1}{kT} \tag{3.7}$$

3.5 The statistical entropy

In this section we establish the link between the statistical weight, Ω, of the predominant configuration and the thermodynamic entropy, S. This link takes the form

$$S = k \ln \Omega \tag{3.8}$$

This expression, first proposed by Ludwig Boltzmann in 1896, is considered again more fully in Chapter 7, where the link between the thermodynamic entropy and the partition function is finally established.

The Boltzmann expression above can be approached starting from eqns 1.6 and 3.4, giving

$$U = U(0) + E = U(0) + \sum_{\text{states}} n_i \varepsilon_i \tag{3.9}$$

We can establish how a small change, dU, in the internal energy can arise through a change in the parameters of eqn 3.9 by noting that U is expressed as a function of two variables (n_i and ε_i; $U(0)$ is a constant) so we can write the overall change, dU, in the internal energy U in terms of small changes, dε_i and dn_i, in these two variables.

$$dU = \sum_{states} n_i d\varepsilon_i + \sum_{states} \varepsilon_i dn_i \tag{3.10}$$

where the first term in the expression, $\sum n_i d\varepsilon_i$, represents the change in U that results from work done on the system, and the second term, $\sum \varepsilon_i dn_i$, represents the change in U that results from the heat taken in by the system.

The internal energy of a perfect gas does not depend on the volume. Hence $d\varepsilon_i = 0$.

One of the terms in eqn 3.10 can be eliminated immediately by noting that, in a perfect gas, the spacing between successive energy states (ε_i) does not alter on heating and remains constant at **constant volume**, which is a defining condition for the internal energy. So the change in energy level spacing, dε_i, is zero and

$$dU = \sum_{states} \varepsilon_i dn_i \tag{3.11}$$

$dU = dq + dw$ (*First Law*)
but, at constant volume, $dw = 0$ and, for a reversible process, $dq = dq_{rev}$.
So $dU = dq_{rev}$
$dS = \frac{dq_{rev}}{T}$ (*Second Law*)
whence, $dU = dq_{rev} = TdS$

From classical thermodynamics we know that an infinitesimal **reversible** change in a system which involves an infinitesimal flow of heat, dq_{rev}, produces change in entropy, dS. If the system remains at **constant volume**, then this change in entropy is related to the change in internal energy by $dU = dq_{rev} = TdS$

so that

$$dS = \frac{dU}{T} = k\sum_{states} \beta\varepsilon_i dn_i \tag{3.12}$$

The condition for a maximum in Ω, which gives the predominant configuration, is that

$$\left(\frac{\partial \ln \Omega}{\partial n_i}\right) = \beta\varepsilon_i - \alpha$$

This, in turn, gives for the change in entropy, dS

$$dS = k\sum_i \left(\frac{\partial \ln \Omega}{\partial n_i}\right) dn_i + k\alpha\sum_i dn_i \tag{3.13}$$

and, since the number of particles is constant ($\Sigma dn_i = 0$) eqn 3.13 leads to $dS = k\,d(\ln \Omega)$ which gives support to the Boltzmann expression

$$S = k \ln \Omega \tag{3.8}$$

3.6 Conclusions

This chapter has established three important points:

- the link between the molecular partition function and thermodynamics,
- the relationship $\beta = 1/kT$, and
- the link between statistical and thermodynamic entropy, $S = k\ln\Omega$.

4 From molecule to mole: the canonical partition function

4.1 Introduction

The molecular partition function contains all the thermodynamic information about a system of independent particles at equilibrium. It also tells us about the number of energy states that are accessible to the system at the temperature of interest.

Independent particles: particles that interact only at collision and thus have no mutual potential energy resulting from their mutual separations.

From eqn 3.2, we can find the **mean molecular energy**, $\langle \varepsilon \rangle$, the average energy per molecule

$$\langle \varepsilon \rangle = \frac{E}{N} = -N\left(\frac{\partial \ln q}{\partial \beta}\right)_V \tag{4.1}$$

from which we can calculate the **molar energy**, E_m, as

$$E_m = -L\left(\frac{\partial \ln q}{\partial \beta}\right)_V \tag{4.2}$$

In going from the molecular to the molar level, we have assumed that the value of an extensive function for N particles is just N times that for a single particle. This is a reasonable assumption for the energy of non-interacting particles, but is not unequivocally so for other properties such as, say, the entropy.

The assumption that particles are **independent** (i.e. non-interacting) is a necessary corollary of writing eqn 4.2, but its necessity is restrictive. We need to allow for the possibility of interactions, and we do so by invoking the idea that every system has a set of **system energy states** which molecules can populate, and that these states are not restricted by the need for additivity but can adjust to any inter-particle interactions that may exist. This chapter explores the consequences of this idea.

4.2 System energy states

Consider a system containing N particles. In order to seek out the most probable **'N-particle' configuration** (or **molar configuration** for one mole of particles) of these, we need to consider them all at once when computing a **molar sum over states**, because each possible state of the whole system involves a description of the conditions experienced by all the particles that make up a mole.

If $N = L$, we are dealing with **molar quantities** and can refer to a **molar configuration**.

This is best illustrated with a concrete yet simple example. Suppose that we have N identical particles, each with the set of individual molecular states available to them.

For our convenience, in order to keep track of what is going on, we use labels. The particles are labelled by number and the molecular states they occupy are labelled by letter

| Particle labels | 1, 2, 3, 4, ..., N |
| Molecular states | $a, b, c, d, ...$ |

Any given **molar state** can be described by a suitable combination of individual molecular states occupied by individual molecules. If we call the i th state Ψ_i, we can begin to give a description of this molar state by writing

$$\Psi_i = 1_a \; 2_b \; 3_h \; 4_f \; 5_c \; 6_k \; 7_c \; 8_t \; ...$$

with energy $\quad E_i = \varepsilon_a^1 + \varepsilon_b^2 + \varepsilon_h^3 + \varepsilon_f^4 + \varepsilon_c^5 + \varepsilon_k^6 + \varepsilon_c^7 + \varepsilon_t^8 \; ...$

Note that, in our example, particles 5 and 7 are both in molecular state c as, of course, is quite proper. There is no restriction on the number of particles that can be in the same molecular state.

State Ψ_i, with energy E_i, is but one of many possible states of the **whole system**. We can now follow exactly the same line of reasoning as we did in Chapter 1, to arrive at an analogue of the predominant molecular configuration (eqn 1.7). This is called the **canonical distribution** and applies to states of an N-particle system, taken as a whole, under conditions of constant amount, volume, and temperature. From the canonical distribution, we can derive a defining equation for the N-particle **canonical partition function**, Q_N, just as we did in eqn 2.6 for its analogue, the molecular partition function, q.

$$Q_N = \sum_{\substack{system \\ states}} e^{-\beta \varepsilon_i} \tag{4.3}$$

If we now average over all possible energies of all possible molar states then, following eqn 2.6, we can write

$$E_m = \langle E_i \rangle = -\left(\frac{\partial \ln Q_N}{\partial \beta} \right)_V \tag{4.4}$$

Equation 4.4 provides us with a way of determining the total energy of an assembly of N-particles ($n = L$ for one mole) considered as a whole, and not as the product of individually identical parts. The canonical partition function, Q_N, is much more general than "the product of N molecular partition functions q," since there is now no need to consider only independent molecules.

The term **canonical** means "according to a rule, or canon". There are several different sets of rules that can be formulated, each giving rise to a different distribution. These different sets of rules describe a set of different *ensembles*, a term that simply means "a collection of particles subject to certain pre-determined rules".
The three main canons (sets of rules) are:

Microcanonical: N, V, E constant
Canonical: N, V, T constant
Grand canonical: μ, V, T constant

where μ is the chemical potential of the particles in the ensemble.

4.3 The molar energy

Although eqn 4.4 is similar to eqn 3.2

$$E = -\frac{N}{q}\left(\frac{\partial q}{\partial \beta}\right)_V = -N\left(\frac{\partial \ln q}{\partial \beta}\right)_V \tag{3.2}$$

it differs in one important respect. Equation 3.2, using the molecular partition function, q, includes a multiplying factor, N, thereby implying the strict scaling of properties from one size of system to another. This can only be true if the particles are truly independent. Equation 4.4, which uses the canonical partition function, Q, has no multiplying factor, N, simply because the N particles have already been considered together, all at once, in its formulation. This lies at the core of the difference between q and Q_N. What is more, it also points to a link between these two partition functions. If we are able to calculate thermodynamic properties for assemblies of N **independent** particles using q, and for N **non-independent** particles using Q_N then, in the limit of the particles of Q_N, becoming less and less strongly interdependent, the two methods should eventually converge.

We note, without the force of it having been proved, that the expressions

$$-N\left(\frac{\partial \ln q}{\partial \beta}\right)_V \quad \text{and} \quad -\left(\frac{\partial \ln Q_N}{\partial \beta}\right)_V$$

are entirely compatible if we assume that the two different partition functions are related simply by the expression

$$Q = q^N \tag{4.5}$$

where, by dropping the subscript N from the canonical partition function, little has been lost since Q is always recognised as a function of an N-particle assembly under conditions of constant temperature and volume.

4.4 Conclusions

In establishing a new partition function, the canonical partition function, Q, which does not rely on the premise that individual particles may not interact with each other, we have taken a major step towards providing a new weapon with which to tackle increasingly complex situations. In suggesting eqn 4.5 as a basis for understanding the relationship between the molecular partition function and the canonical partition function, this step has been enhanced. In the next chapter we shall establish the importance of eqn 4.5 and explore both its limitations and its uses.

5 Distinguishable and indistinguishable particles

5.1 Introduction

In eqn 4.5, we suggested that a link, or at least a feasible link, could be forged between the molecular partition function, q, and the canonical partition function, Q. In this chapter we shall seek to confirm this link, to show the conditions under which it is valid, and to explore alternative expressions when it is not.

5.2 Linking *q* to *Q*

Under the conditions where it is valid to use either of these partition functions, Q and q (and these occur when there are no bulk interactions between particles to contribute to a potential energy term), both approaches must yield the same molar energy. In the example used in Chapter 4, where we wrote the energy of the i th system state as

$$E_i = \overset{1}{\varepsilon}_a + \overset{2}{\varepsilon}_b + \overset{3}{\varepsilon}_h + \overset{4}{\varepsilon}_f + \overset{5}{\varepsilon}_c + \overset{6}{\varepsilon}_k + \overset{7}{\varepsilon}_c + \overset{8}{\varepsilon}_t \ \ldots$$

for the canonical partition function we can write

$$Q = \sum_i \exp[-\beta(\overset{1}{\varepsilon}_a + \overset{2}{\varepsilon}_b + \overset{3}{\varepsilon}_h + \overset{4}{\varepsilon}_f + \overset{5}{\varepsilon}_c + \overset{6}{\varepsilon}_k + \overset{7}{\varepsilon}_c + \overset{8}{\varepsilon}_t \ \ldots)]$$

Now, in every one of the i system states, each particle (1, 2. 3, ...) will be in one of its possible, j, molecular states (a, b, c, d, ...), just once (and once only) in each system state. So, if we extract (factorise out) each particle in turn from the summation over the system states in eqn 5.1 and then gather together all the terms that refer to a given particle, we get

Remember that $e^{-(a+b)} = e^{-a}e^{-b}$

It is worth trying to do this using only two particles, *a* and *b*, for which it is simple to show that

$$\sum e^{-\beta(\varepsilon_a+\varepsilon_b)} = \sum e^{-\beta\varepsilon_a} \times \sum e^{-\beta\varepsilon_b}$$

$$Q = \left(\sum_{\substack{molecular\\states}} e^{-\beta\varepsilon_j}\right)_1 \left(\sum_{\substack{molecular\\states}} e^{-\beta\varepsilon_j}\right)_2 \left(\sum_{\substack{molecular\\states}} e^{-\beta\varepsilon_j}\right)_3 \ldots$$

and, if all the molecules are of the same type and indistinguishable by position they do not need labelling, so

$$Q = \left(\sum_j e^{-\beta\varepsilon_j}\right)^N = q^N \qquad (5.2)$$

Equation 5.2 provides proof for the intuitive assignment we made in eqn 4.5. The link between the molecular partition function and the canonical partition function is now established.

5.3 Distinguishable and indistinguishable particles

The relationship between the canonical and molecular partition functions in eqn 5.2 is correct if the particles in the assembly, though identical in type, have distinct and recognisable identities. If they have some recognisable attribute or label that distinguishes them uniquely from each of their fellows, then we say they are **distinguishable** and eqn 5.2 holds.

If particles are not distinguishable in this sense, then eqn 5.2 overcounts the total number of system states present. If particles are **indistinguishable**, the number of accessible system states is lower (often drastically lower) than it is for distinguishable ones. We can illustrate this as follows.

A system state, Ψ_i, for which the configuration includes the assignment of particles to states

$$\Psi_i = 1_a\, 2_b\, 3_h \cdots$$

differs from a state Ψ_j

$$\Psi_i = 1_a\, 2_h\, 3_b \cdots$$

$\Psi_i \equiv \Psi_j$ for indistinguishable particles
$\Psi_i \neq \Psi_j$ for distinguishable particles

if particles are **distinguishable**, because of the interchange of particles 2 and 3 between states b and h. In contrast, however, state, Ψ_i is **identical** to state Ψ_j if particles are **indistinguishable**. Thus

$Q = q^N$ overestimates Q for **indistinguishable** particles.

In systems which are not at too high a density and are also well above 0 K, the correction factor for this overcounting of configurations is $1/N!$ So

$$Q = q^N \quad \text{for distinguishable particles}$$

$$Q = \frac{q^N}{N!} \quad \text{for indistinguishable particles} \qquad (5.3)$$

We shall show this by inference in the next section, and then address the question of what it is that confers **distinguishability** on atomic or molecular particles.

5.4 The origin of 1/N!

Consider just two particles, 1 and 2, in an assembly. These are free to occupy any of the a, b, c, d, ... molecular states of the system. Since each state is weighted by a Boltzmann factor involving the energy of that state, in the canonical partition function we will find terms such as

$$e^{-\varepsilon_a^1} + e^{-\varepsilon_b^2} \text{ or, in shorthand, } 1a2b$$

where the exponential term is identified both by the label of the molecule (1, 2) and by the energy of the state ($\varepsilon_a = a$, $\varepsilon_b = b$, ..., etc). If we pick out from the assembly all the terms involving just particle 1 and particle 2, and arrange these in a systematic manner, we obtain the following matrix

	1*a*2*b*	1*a*2*c*	1*a*2*d*	1*a*2*e*	...
1*b*2*a*		1*b*2*c*	1*b*2*d*	1*b*2*e*	...
1*c*2*a*	1*c*2*b*		1*c*2*d*	1*c*2*e*	...
1*d*2*a*	1*d*2*b*	1*d*2*c*		1*d*2*e*	...
1*e*2*a*	1*e*2*b*	1*e*2*c*	1*e*2*d*		...
⋮	⋮	⋮	⋮	⋮	

If particles 1 and 2 are *distinguishable*, then each of the pairs above represents a distinct contribution to a configuration. But, if they are *indistinguishable*, then, for instance, 1*a*2*b* is the same as 1*b*2*a* and, since the matrix is symmetrical about its diagonal, overcounting by a factor of 2 results. The diagonal elements are left blank because the analysis, as we shall see shortly, depends on each particle occupying a *separate* molecular state.

There are three ways of picking the first particle (to occupy state *a*), then two ways of picking the second (to occupy state *b*). The third particle picks itself (one way) to occupy state *c*. Total ways = 3 × 2 × 1 = 6.

$\frac{1}{2} = \frac{1}{2!}$

$\frac{1}{6} = \frac{1}{3!}$

$\frac{1}{24} = \frac{1}{4!}$

In the case of two indistinguishable particles, then, we need to introduce a correction factor of 1/2. How do we extend this to three particles and more? Three particles treated along the lines above produce a three-dimensional matrix; N particles require N dimensions. But we need not go to such extremes to detect the trend. For three particles, 1, 2, and 3, occupying only three states, a, b, and c, there are **six** distinct ways of shuffling the particles among the states if the particles are labelled (i.e. *distinguishable*) but only **one** if they are not (i.e. *indistinguishable*). For four particles and four states, the ratio of *distinguishable* to *indistinguishable* arrangements is 24. For N particles occupying N states, the factor is $N!$.

5.5 The number of states per particle

The analysis above is only valid if the N particles occupy N essentially distinct quantum states. In turn, this requires that the number of quantum states accessible to each particle is large, otherwise several will occupy the same state. The accessibility of states is reflected by the magnitude of q, so the requirement for the valid application of a correction factor of $1/N!$ is

$$\frac{q}{L} \gg 1$$

requiring that the number of accessible states should greatly exceed the total number of particles.

At room temperature q is typically of the order of 10^{30}, so that $q/L \sim 10^7$.

For translational motion, $q \gg L$ at most reasonable temperatures; at very low temperatures, the accessibility of quantum states collapses dramatically, so that q/L no longer exceeds unity. In such cases it may become necessary to abandon the statistics we have used so far (called **Maxwell–Boltzmann statistics**) in favour of **quantum statistics** (known as **Bose–Einstein**

Maxwell–Boltzmann statistics
Bose–Einstein statistics
Fermi–Dirac statistics

statistics or **Fermi-Dirac statistics**). This topic lies beyond the scope of this book, but it is as well to be aware of its existence.

The only gas whose translational motion is of importance near 0 K is helium; for most gases, $q/L \gg 1$, so it is perfectly safe to write

$$Q = q^N \quad \text{for distinguishable particles}$$
$$Q = \frac{q^N}{N!} \quad \text{for indistinguishable particles} \tag{5.3}$$

5.6 What is indistinguishability?

Why do we need to categorise some particles as distinguishable, and others as indistinguishable? To what properties of the particles can these characteristics be attributed?

Clearly, particles of different types are *intrinsically distinguishable*. Helium atoms can never masquerade as xenon atoms; they are much too light! So mass will always provide a distinguishing feature between particles. But what about distinguishing between individual xenon atoms?

If the particles in question are in the solid state and form part of a crystal lattice then, by virtue of this, they too are *intrinsically distinguishable*. Each particle in a crystal lattice has an individual three-dimensional address (e.g. 4th row, 7th column, 12th atom in) which will always distinguish it from any other particle. So atoms in solid xenon are *always distinguishable* and can never lose their identities. For them, $Q = q^N$.

But if the particles are in the gas phase, they are free to move around from place to place and there is no way, even in principle, of keeping track of where they are at any instant. The reason for this is fundamental to the nature of quantum behaviour. In a classical collision, even between two identical white billiard balls, we can, in principle, keep track of the trajectories of the two colliding partners. In a collision between atoms we cannot, at least not precisely. An ability to do so would violate Heisenberg's uncertainty principle which leads us to the conclusion that identical quantum particles in the gas phase must be *intrinsically indistinguishable* since we cannot follow their tracks with complete certainty through even the most simple of collisions. So gas particles are *always indistinguishable*. For them, $Q = q^N/N!$, as long as $q/L \gg 1$.

In between, for liquids and vapours, lies an unknown hinterland. It is of great interest theoretically and practically but, unfortunately, well outside the scope of this introductory text.

5.7 Conclusions

In this chapter we have looked at the concept of indistinguishability at the microscopic level and have suggested how this can be incorporated into our master equations for the canonical partition function. We now have an unequivocal set of relationships linking the microscopic and the macroscopic worlds. The rest of this book sets out to exploit this understanding.

6 Two-level systems: a case study

6.1 Introduction

We are now in a position to apply the ideas developed in earlier chapters to real systems. The simplest type of system to fall into this category is one which comprises particles with only two accessible states in the form of two non-degenerate levels separated by a narrow energy gap $\Delta\varepsilon$. This is shown schematically in Fig. 6.1, and is discussed in Section 6.2 below.

Such systems are not entirely hypothetical. Real examples include protons which are distributed over two closely spaced energy states in a magnetic field (as in NMR), and atoms or molecules with a single, low-lying, electronic excited state (such as cerium(III) ions in certain salts, or nitrogen monoxide molecules in the gas phase).

In a certain sense, all energy levels at low enough temperatures eventually approximate to two-state systems (and finally to featureless one-level systems) as the temperature drops. At temperatures that are comparable to $\Delta\varepsilon/k$, only the ground state and the first excited state are appreciably populated, a feature that leads to the unexpected rotational heat capacity behaviour of *ortho-* and *para-*hydrogen below 100 K, and to the characteristic vibrational heat capacity behaviour of hydrogen at around 4000 K.

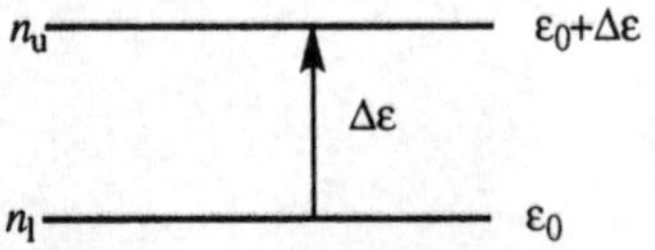

Fig. 6.1 A two-level system. The lower level, with energy ε_0 has a population n_l and is separated by an energy gap $\Delta\varepsilon$ from the upper level, with energy $\varepsilon_0 + \Delta\varepsilon$ and a population n_u.

6.2 The effect of increasing temperature

Characteristic temperatures

Consider a system with two non-degenerate levels separated by an energy gap $\Delta\varepsilon$ as shown schematically in Fig. 6.1. Applying the Boltzmann law, the average population ratio of the two levels in an assembly of such two-level (or two-state) particles is given by eqn 6.1.

$$\frac{n_u}{n_l} = e^{-\beta\Delta\varepsilon} = e^{-\theta_{2L}/T} \tag{6.1}$$

where the characteristic two-level temperature, θ_{2L}, is defined as

$$\theta_{2L} = \frac{\Delta\varepsilon}{k} \tag{6.2}$$

Some characteristic temperatures are given in Table 6.1. Those for hydrogen are, of course, not characteristic two-level temperatures since, in rotation and vibration, hydrogen has manifolds of states between which gaps are not constant. The other three, however, can be considered as true two-level cases.

Table 6.1 Selected characteristic temperatures, θ

System		θ/K
Proton – 10 T field	(mag)	0.02
Cerium(III)	(el)	7
Hydrogen	(rot)	90
Nitrogen monoxide	(el)	178
Hydrogen	(vib)	4440

mag = magnetic, el = electronic, rot = rotational, vib = vibrational

Temperature dependence of the populations

Deciding what is a high or a low temperature is best approached by comparing the energy gap, $\Delta\varepsilon$, with the background thermal energy, kT, at the temperature in question. More conveniently, we can compare the characteristic temperature, θ_{2L} (eqn 6.2) directly with the temperature T.

$\frac{\Delta\varepsilon}{kT} \equiv \frac{\theta}{T}$

Now, if $T = 5 \times \theta_{2L}$ (a relatively high temperature), the population ratio n_u/n_l is quite high (0.82) and is reasonably close to unity. Indeed, the higher the temperature (relative to θ_{2L}), the closer to unity the ratio becomes. But if $T = 0.5 \times \theta_{2L}$ (a relatively low temperature), the population ratio n_u/n_l is quite low (0.14) and is reasonably close to zero. Indeed, the lower the temperature (relative to θ_{2L}), the closer to zero the ratio becomes.

If $T = 5\theta_{2L}$ (high T)
then $\frac{n_u}{n_l} = e^{-1/5} = 0.82$ (or ≈ 1)

If $T = 0.5\theta_{2L}$ (low T)
then $\frac{n_u}{n_l} = e^{-1/0.5} = 0.14$ (or ≈ 0)

Rather than deal with ratios, it makes good sense to derive values for the upper and lower level populations directly. Starting with eqn 6.1, n_l can be expressed in terms of n_u

$$n_l = e^{\beta\Delta\varepsilon} n_u \tag{6.3}$$

but, since the total number of particles is constant,

$n_l + n_u = N$

so

$$\left(1 + e^{\beta\Delta\varepsilon}\right) n_u = N \tag{6.4}$$

$$n_u = \left(\frac{1}{1+e^{\beta\Delta\varepsilon}}\right) N = \left(\frac{e^{-\beta\Delta\varepsilon}}{1+e^{-\beta\Delta\varepsilon}}\right) N \tag{6.5}$$

and

$$n_l = \left(\frac{e^{\beta\Delta\varepsilon}}{1+e^{\beta\Delta\varepsilon}}\right) N = \left(\frac{1}{1+e^{-\beta\Delta\varepsilon}}\right) N$$

Each of the right-hand quotients in eqns 6.5 are obtained by multiplying both numerator and denominator in the left-hand quotients by $e^{-\beta\Delta\varepsilon}$.

In Fig. 6.2, the arguments of eqns 6.5 are plotted as functions of T/θ_{2L}, the ratio of the actual temperature to the characteristic temperature. This ratio of these two temperatures is called a **reduced temperature.** It is a dimensionless ratio, directly proportional to the actual temperature but specifically scaled to the actual energy gap of the system in question. Furthermore, in order to display clearly the asymptotic behaviour at very low and very high temperatures (as $T \rightarrow 0$ and as $T \rightarrow \infty$), the reduced temperature has been plotted on a logarithmic scale.

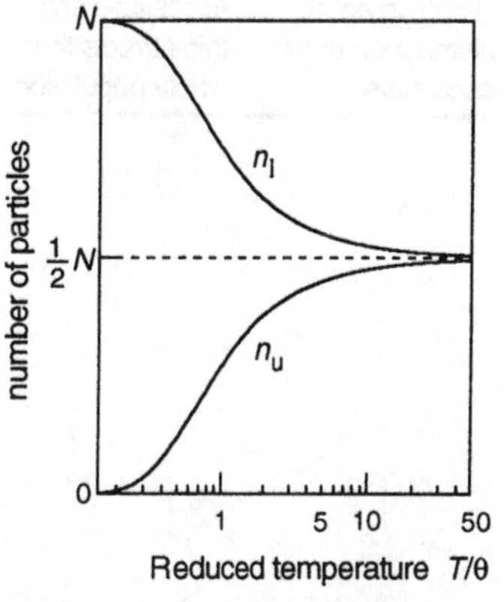

Fig. 6.2 The number of particles in the upper and lower states of a two-state system as a function of the reduced temperature T/θ. Note the logarithmic temperature scale which clarifies what is happening both as $T \rightarrow 0$ and as $T \rightarrow \infty$.

As might be expected, the lower level population starts with N at low temperatures, where the upper level population is zero because there is not enough thermal energy available to raise any particles into the upper state. At high temperatures, the populations are equalised and both tend to the limiting value $\frac{1}{2}N$.

Note that

$$\textit{reduced temperature} = \frac{T}{\theta} = \frac{kT}{\Delta\varepsilon} = \frac{1}{\beta\Delta\varepsilon} \tag{6.6}$$

so that $e^{-\theta/T} = e^{-\beta\Delta\varepsilon}$.

We always have the option of choosing the energy zero with respect to which we decide to scale all our energies because, in those cases where this choice matters, the actual energy of the zero reference is always subsumed into the term representing the internal energy at the absolute zero, $U(0)$. Situations where $U(0) \neq 0$ can arise only if the particles in question have potential energy by virtue of interactions that depend on their separations. Such interactions are specifically excluded in the perfect gas model used throughout this text.

Table 6.2 Variation of the two-level partition function with temperature

$T = 5\theta_{2L}$	$T = 0.5\theta_{2L}$
$q_{2L} = 1 + 0.82$	$q_{2L} = 1 + 0.14$
$= 1.82$	$= 1.14$
So (high T), when	and (low T), when
$T \gg \theta_{2L}$	$T \ll \theta_{2L}$
$q_{2L} \to 2$	$q_{2L} \to 1$
Both states are equally accessible giving an equal distribution over each state	Only one state (the lower) is accessible and this attracts the whole population

6.3 The two-level molecular partition function

The effect of increasing temperature

With only two energy states to consider, we can easily write an expression for the two-level molecular partition function, q_{2L}.

$$q_{2L} = \sum_{\substack{both\\states}} e^{-\beta\varepsilon_i} = e^{-\beta\varepsilon_0} + e^{-\beta(\varepsilon_0+\Delta\varepsilon)} = e^{-\beta\varepsilon_0}\left(1 + e^{\beta\Delta\varepsilon}\right)$$

or, setting $\varepsilon_0 = 0$, following our convention from Chapter 1,

$$q_{2L} = \left(1 + e^{-\beta\Delta\varepsilon}\right) \tag{6.7}$$

The effect of temperature on the value of the two-level partition function is summarised in Table 6.2. When both states are equally accessible, (at high temperatures) the molecular partition function approaches its limiting maximum value of 2. The (low temperature) limiting value is 1.

6.4 The energy of a two-level system

Simple summation leads to an expression for the energy of a two-level system. Once again, we set the ground-state energy to zero for convenience.

$$E_{2L} = \sum_i n_i\varepsilon_i = n_l \times 0 + n_u \times \Delta\varepsilon$$

$$= \left(\frac{N\Delta\varepsilon}{1 + e^{\beta\Delta\varepsilon}}\right) \tag{6.8}$$

which is obtained by substituting the expression for n_u from eqn 6.5.

It is instructive to see how the same expression could have been obtained using the two-level partition function, eqn 6.7. We start by recalling the equation that links the energy of an N-particle system to the temperature dependence of the molecular partition function, as was done in eqn 3.2, and then modified to include constancy of volume in Section 4.3.

$$E = -\frac{N}{q}\left(\frac{\partial q}{\partial\beta}\right)_V = -N\left(\frac{\partial \ln q}{\partial\beta}\right)_V \tag{3.2}$$

which, because the two-level partition function does not depend on the volume of the system, can be written without partial notation or constancy of volume, as in Section 3.2.

$$E = -N\left(\frac{d\ln q}{d\beta}\right) = -\frac{N}{q}\left(\frac{dq}{d\beta}\right) \tag{6.9}$$

$$d\ln x = \frac{dx}{x}$$

The expression on the right-hand side of eqn 6.9 follows from the properties of differentials, *viz.* $d\ln x = dx/x$.

Differentiating eqn 6.7, we get

$$\left(\frac{dq}{d\beta}\right) = -\Delta\varepsilon\, e^{-\beta\Delta\varepsilon} \qquad (6.10)$$

whence

$$E = \left(\frac{N\Delta\varepsilon\, e^{-\beta\Delta\varepsilon}}{1+e^{-\beta\Delta\varepsilon}}\right) = n_u\Delta\varepsilon \qquad (6.11)$$

a result identical to the expression in eqn 6.8 if both numerator and denominator above are multiplied by $e^{+\beta\Delta\varepsilon}$.

The general shape of the temperature dependence of the total energy of a two-level system follows exactly the trace of the lower curve in Fig. 6.2, but with the vertical axis scaled by an additional factor of $\Delta\varepsilon$. As might be expected, the total energy starts at zero at low temperatures since, at low temperatures, the upper level population is zero because there is not enough thermal energy available at low temperatures to raise many particles into the upper state. At high temperatures, half the particles occupy the upper state and the total energy takes the value $\frac{1}{2}N\Delta\varepsilon$.

The total energy of a two-level system depends entirely on the population of the upper state and thus looks exactly like the lower curve in Fig. 6.2.

6.5 The two-level heat capacity, C_V

The heat capacity at constant volume

The spacing of the energy levels to which we refer in discussing two-level systems is not affected by changes in volume, so the relevant heat capacity is C_V not C_P (see also Section 7.8).

Qualitative assessment

In general, the heat capacity C_V reflects the accessibility of states above the ground state. For a two-level system there is only one other accessible state, and the variation of the heat capacity with temperature gives a measure of just how accessible this upper state becomes as the temperature is altered.

$$C_V = \left(\frac{\partial U}{\partial T}\right)_V \approx \frac{\Delta U}{\Delta T}$$

At low temperatures, most particles have little tendency to leave the lower level. The thermal energy, kT, is rather small, so a small change in temperature, ΔT, has little tendency to excite particles into the upper level. The overall energy of the system remains roughly constant, so the heat capacity, C_V, is rather low.

At intermediate temperatures, kT becomes comparable to $\Delta\varepsilon$ and the number of particles in the upper level starts to rise. A small change in temperature now has a much larger effect on the population in the upper state, so the heat capacity is somewhat higher.

At high temperatures, the system is close to *saturation*, with almost half the particles in the excited state. A small change in temperature causes very few extra particles to be excited into the upper level. The overall energy of the system remains roughly constant, so the heat capacity, C_V, is once again rather low.

So we expect the heat capacity of a two-level system to start near zero at very low temperatures, to rise and pass through a maximum as the temperature increases, and then to decline to zero again at very high temperatures. Such behaviour was predicted theoretically in 1922 by the physicist W. Schottky. This unusual behaviour of the heat capacity is called a **Schottky anomaly**.

Schottky anomaly

Quantitative predictions

In order to obtain the two-level heat capacity, we have to differentiate the energy eqn 6.11 with respect to temperature. Since the energy, E, is a function of temperature only and not of volume, the partial nature of the derivative can be dropped.

Recall that $\beta = \frac{1}{kT}$

Now, from eqn 6.11,
$$E = N\Delta\varepsilon\left(1+e^{\beta\Delta\varepsilon}\right)^{-1} \quad (6.12)$$

so
$$\frac{dE}{dT} = -N\Delta\varepsilon\left(1+e^{\beta\Delta\varepsilon}\right)^{-2} e^{\beta\Delta\varepsilon}\left(-\frac{\Delta\varepsilon}{kT^2}\right)$$

$$C_V = Nk(\beta\Delta\varepsilon)^2\left[\frac{e^{\beta\varepsilon\Delta}}{(e^{\beta\varepsilon\Delta}+1)^2}\right] \quad (6.13)$$

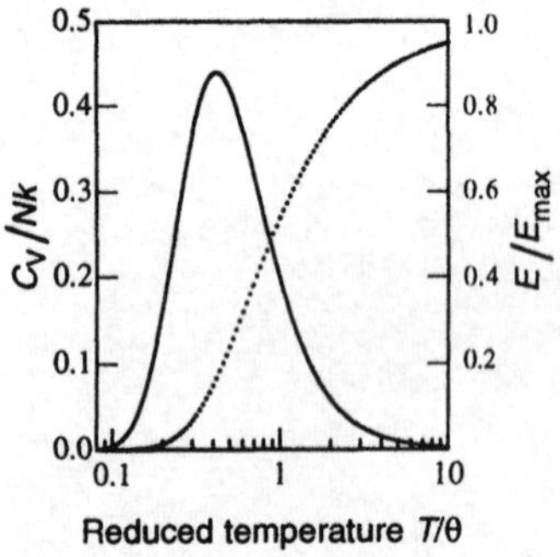

Fig. 6.3 The heat capacity of a two-level system as a function of the reduced temperature. Also shown is the variation of the two-level energy with reduced temperature.

The two-level heat capacity (eqn 6.13) is plotted as a function of temperature in Fig. 6.3; also shown is the energy, E (eqn 6.11). As predicted qualitatively above, the heat capacity does rise from zero at low temperatures to a maximum value before declining again to zero as the temperature rises yet further and the upper and lower populations become equalised. Now, however, we can understand why the maximum value of the heat capacity is $0.44Nk$ and why this maximum occurs at a reduced temperature of 0.42. The logarithmic reduced temperature scale shows these features clearly. The maximum in the $C_V(T)$ graph corresponds to the point of inflection in the $E(T)$ curve.

Comparison with experiment

As mentioned in Section 6.1, real examples of two-level behaviour do exist. The salt cerium(III) ethylsulfate, $Ce(C_2H_5SO_4)_3$, has a heat capacity that shows an anomaly close to the absolute zero. The cerium(III) ion in this salt is known to have a low-lying doublet first excited electronic state, $^2D_{3/2}$ and $^2D_{5/2}$, which lies only about 500 cm^{-1} above the doublet ground electronic state, $^2F_{5/2}$ and $^2F_{7/2}$. This energy separation, as shown in Table 6.1, corresponds to a gap with a characteristic temperature $\theta \approx 7$ K. Careful measurements of the low temperature heat capacity reveal a peak in the heat capacity close to 3 K, as shown in Fig. 6.4. The closeness of theoretical predictions to experiment is most satisfying. In terms of temperature, $0.42 \times 7\,K \approx 2.9\,K$, and in terms of molar heat capacity, $0.44\,R \approx 3\,J\,K^{-1}\,mol^{-1}$, both falling remarkably close to the experimental results. Two-level systems *do* exist.

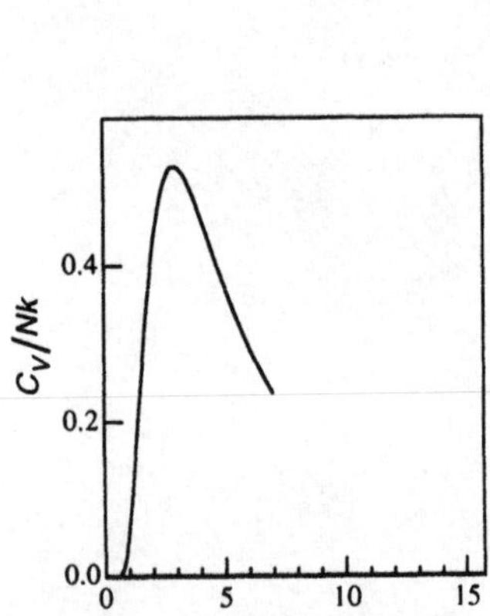

Fig. 6.4 The measured molar heat capacity of cerium ethylsulfate.

Lest it be thought that small anomalies near the absolute zero are somewhat esoteric, it is worth noting that such seemingly minor discrepancies have played an important part in providing experimental evidence for the validity of the Third Law of thermodynamics. Several cases where apparently discordant calorimetric and spectroscopic data seemed destined to upset the validity of the Third Law have been resolved through a better understanding of low temperature Schottky anomalies. Furthermore, the actual magnitude of the anomaly (the height of the C_V peak) is far from negligible. Iron(III) aluminium sulfate exhibits a Schottky anomaly at an even lower temperature (0.05 K) where all other contributions to the heat capacity such as lattice vibrations (or *phonons*) have long since decayed virtually to zero. It has been estimated that one cubic centimetre of iron(III) ammonium sulfate at 0.05 K has the same heat capacity as 16 tonnes of lead at the same temperature. This is by no means a trivial effect!

6.6 The effect of degeneracy

Consider a two-level system with degenerate energy levels. If the degeneracy of the lower level is g_0 and that of the upper level is g_1 then the two-level molecular partition function (*cf.* eqn 6.7) is

$$q_{2L} = \left(g_0 + g_1 e^{-\beta\Delta\varepsilon}\right) \tag{6.14}$$

and, following the procedure adopted for eqn 6.11, we find for the energy of the degenerate two-level system

$$E = \left[\frac{\left(\frac{g_1}{g_0}\right) N\Delta\varepsilon}{\left(\frac{g_1}{g_0} + e^{\beta\Delta\varepsilon}\right)}\right] \tag{6.15}$$

and for the heat capacity

$$\frac{C_V}{Nk} = (\beta\Delta\varepsilon)^2 \left[\frac{\left(\frac{g_1}{g_0}\right) e^{\beta\Delta\varepsilon}}{\left(\frac{g_1}{g_0} + e^{\beta\Delta\varepsilon}\right)^2}\right] \tag{6.16}$$

The family of heat capacity functions given by eqn 6.16 is plotted in Fig. 6.5. It is clear that as the degeneracy ratio becomes larger, the peak in the $C_V(T)$ graph becomes higher and shifts to slightly lower temperatures. As we see in Ch. 14, such effects are encountered at low temperatures in the rotational heat capacity of diatomic molecules such as HD, and also in *o*- and *p*-hydrogen. An understanding of these phenomena can be improved by reference to the behaviour shown in Fig. 6.5.

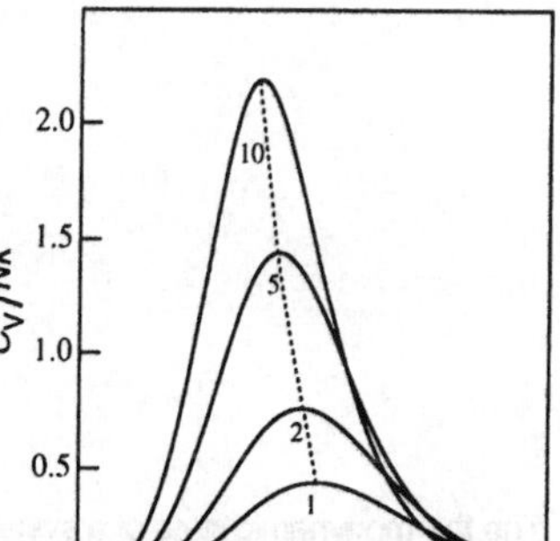

Fig. 6.5 The effect of degeneracy on the anomalous two-level heat capacity. The degeneracy ratio, g_1/g_0 is shown beneath each peak.

6.7 Conclusions

The thermal behaviour of a two-state system can be modelled satisfactorily using partition functions. Theoretical predictions for real systems correspond well with experimental results.

7 Thermodynamic functions: towards a statistical toolkit

7.1 Introduction

In the last chapter, we modelled the behaviour of two-level systems and of their partition functions in a manner that enabled us to describe the temperature dependence of their energy and heat capacity. We were able to do so in a relatively straightforward way because of the essential simplicity of the system under investigation. The objective of this chapter is to generalise this description to cover all possible cases and to develop a formalism that will link the partition function, in turn, to each of the classical thermodynamic functions, which provide a description of the equilibrium behaviour of all matter.

The development of the required formalism is not particularly complicated, but those who simply want to use the resulting relationships might wish to pay less than close attention to the reasoning involved. Indeed, it is with this aspect in mind that this chapter has been subtitled *towards a statistical toolkit* since it is perfectly possible (and not at all disreputable) to treat statistical thermodynamics as a means to an end without following too deeply the mechanism by which this means has been put together. Hence, the suite of equations that this chapter will furnish can be considered as individual tools in a comprehensive toolkit, each designed to make it easier to solve specific problems.

Therefore, Sections 7.2 to 7.4 can be treated as an enhancement and not essential to our understanding and use in subsequent chapters. In Section 7.2, we recall that all thermodynamic variables can be defined completely by just two independent variables, called **proper variables**. In Sections 7.3 and 7.4 we establish the link between the partition function and the internal energy, U, and then the entropy, S.

It is enough, therefore, to recognise and accept the importance of certain key relationships in Section 7.5 onwards for the remainder of this book to become accessible.

The thermodynamic state of a system is specified by the variables **pressure, volume, temperature,** and **amount of substance**. One of these variables is not independent because there exists an **equation of state**, and the amount of substance can be factored out by referring to molar quantities, which are **intensive** rather than **extensive.** Thus we are left with two proper independent variables which express the functional variation of any of the thermodynamic state functions:

$U = U(S,V)$
$H = H(S,p)$
$G = G(p,T)$
$A = A(V,T)$

7.2 State functions

The major thermodynamic variables are **state functions**, that is, the values they assume depend only on the **state** in which they find themselves and not on the route followed in reaching that state. Intensive thermodynamic state variables can generally be expressed as functions of **two** other variables (**proper variables**), some examples of which are shown alongside. The ability completely to define any one thermodynamic variable as a function of only two others means that, if we can express any two thermodynamic variables in terms of the **partition function**, all the remaining thermodynamic variables can then be derived from just two others. The two thermodynamic functions most usually chosen as the basis for all the others are the energy E (leading to the internal energy U) and the entropy S.

7.3 The internal energy, U

It is a relatively straightforward matter to calculate the excess of energy, E, above the state of zero energy, $\varepsilon_0 = 0$, that is possessed by an assembly of independent particles. This energy is given by the sum of products such as $n_i\varepsilon_i$, the number of particles at each energy level multiplied by the value of that energy,

$$E = \sum_{\text{states},\, i} n_i\varepsilon_i \tag{7.1}$$

In carrying out the summation in eqn 7.1, we sum from the lowest state ($n = n_0$, $\varepsilon_0 = 0$). The total sum thus achieved represents the energy that the assembly has over and above its energy when in the lowest state ($n = n_0$, $\varepsilon_0 = 0$). In order to obtain from this the internal energy, U, we need to take account of whatever energy, $U(0)$, the system possesses at the absolute zero, when only the lowest state is populated. Thus

$$E = U - U(0)$$

and

$$U - U(0) = \sum_{\text{states},\, i} n_i\varepsilon_i \tag{7.2}$$

Had we chosen some other energy zero, the value of $U(0)$ would have changed to accommodate this.

The summation in eqn 7.2 has effectively been evaluated in Chapter 3, yielding the expression

$$U - U(0) = NkT^2\left(\frac{\partial \ln q}{\partial T}\right)_V \tag{3.5}$$

and, since

$$N\left(\frac{\partial \ln q}{\partial T}\right)_V = \left(\frac{\partial \ln Q}{\partial T}\right)_V$$

whether the link between the canonical partition function, Q, and the molecular partition function, q, is $Q = q^N$ (distinguishable) or $Q = q^N/N!$ (indistinguishable),

$$U - U(0) = kT^2\left(\frac{\partial \ln Q}{\partial T}\right)_V \tag{7.3}$$

We shall use eqn 7.3, an expression involving the **canonical partition function**, Q, in preference to that involving the molecular partition

Recall that for a perfect gas at 0 K, the internal energy is forced to be zero, so the definition $\varepsilon_0 = 0$ is necessarily correct and complete and $U(0) = 0$. However, the perfect gas at 0 K is a hypothetical state, and real systems will experience interactions that lead to a potential energy that is not zero at 0 K, so that $U(0) \neq 0$.

$\ln q^N/N! = \ln q^N - \ln N!$ so $\ln q^N/N! = \ln q^N$ minus a term independent of T.

Hence the temperature derivatives of $N\ln q$ and $\ln Q$ are always identical.

function, q (eqn 3.5), because of the more general applicability of the former to all cases, distinguishable and indistinguishable.

Equation 7.3 is the first of our two seminal equations needed as a route to all other thermodynamic quantities.

7.4 The entropy, *S*

Next, we derive an expression for the entropy, S, as our choice of our second necessary equation. The expression that results can be derived by a number of different means, most notably by using one of the most celebrated equations in modern science,

$$S = k \ln \Omega \tag{7.4}$$

or, as is nowadays becoming common, $S = k \ln W$. The less modern, but more historically correct, form of this equation, as expressed in eqn 7.4 and used throughout this text, forms the moving epitaph on the grave in Vienna of Ludwig Boltzmann, a prime originator of all that concerns this book.

The approach involving eqn 7.4 is the chosen route in many texts, but it is a route that demands of the reader a much deeper understanding of the philosophy and implications behind the meaning of entropy than a brief introduction of this kind can provide. The reader is encouraged to follow this route independently, not only to feel comfortable with the development outlined below, but also to savour the full flavour of developments at the turn of the present century which have had so profound an influence on the whole of our view of science.

Here, instead, we approach the definition of entropy from the more familiar *thermal* viewpoint which is likely to be more transparent to all who have acquired a grounding in classical thermodynamics but who come fresh to statistical thermodynamics.

We start, then, with the derivation of the heat capacity from the energy (eqn 7.3). The route we shall follow invokes, first, the classical relationship

dq_{rev} is the energy supplied reversibly to a system as heat.. That the same symbol is used both for *heat* and for the molecular partition function is regrettable, but unavoidable. Fortunately, the likelihood of any confusion is small.

$$\mathrm{d}S = \frac{\mathrm{d}q_{rev}}{T}$$

from which we get

$$\mathrm{d}S = C_V \frac{\mathrm{d}T}{T} \tag{7.5}$$

and, secondly, the first derivative with respect to temperature, of eqn 7.3.

$$C_V = \left\{\frac{\partial(U - U(0))}{\partial T}\right\}_V = \left\{\frac{\partial}{\partial T}\left[kT^2\left(\frac{\partial \ln Q}{\partial T}\right)\right]_V\right\}_V \tag{7.6}$$

$$= 2kT\left(\frac{\partial \ln Q}{\partial T}\right)_V + kT^2\left(\frac{\partial^2 \ln Q}{\partial T^2}\right)_V$$

Equations 7.5 and 7.6 lead to

$$S = k \ln Q + kT\left(\frac{\partial \ln Q}{\partial T}\right)_V + S_0 \tag{7.7}$$

Since S_0, the constant of integration, is zero (**Third Law**), from eqn 7.3 we reach the conclusion

Third Law: $S(0) = 0$ for perfect crystals can also be applied in the case of the perfect gas, as here.

$$S = \frac{(U - U(0))}{T} + k \ln Q \qquad (7.8)$$

Equation 7.8 for the entropy, our second seminal equation, can be compared directly with the classical expression

$$A = U - TS \qquad (7.9)$$

This expression for the Helmholtz free energy, as was true for the internal energy, gives the difference between the value of A at the absolute zero, $A(0)$, when only the lowest level ($\varepsilon_0 = 0$) is populated, and its value at any other temperature, T. Consequently, eqn 7.9 can more fully be written as

$$A - A(0) = U - U(0) - TS \qquad (7.10)$$

an equation that immediately implies that $A(0) = U(0)$. Rearranging, this gives

$$S = \frac{(U - U(0))}{T} - \frac{(A - A(0))}{T}$$

Comparison with eqn 7.8 now yields an expression for the Helmholtz free energy, A,

$$A - A(0) = -kT \ln Q \qquad (7.11)$$

It is this expression that we shall use in order to determine all the thermodynamic properties using the partition function.

7.5 The Helmholtz energy, *A*: the Massieu bridge

The relationship

$$J = -\frac{A}{T}$$

is known as the **Massieu function**. So, combining this with eqn 7.11, we get an equation which can be called the **Massieu bridge**

$$J = -\frac{(A - A(0))}{T} = k \ln Q \qquad (7.12)$$

THE MASSIEU BRIDGE

The Massieu bridge: a link to thermodynamics.

which will form the basis of all the other thermodynamic functions that we need. It forms a succinct and compact link between statistical and classical thermodynamics and is often called **the link to thermodynamics.** We show how it is used in the sections that follow.

7.6 The internal energy, *U*, revisited

We can use the Massieu bridge to reconstitute the expression for the internal energy. First of all we recall that

$$U = T^2\left(\frac{\partial(A/T)}{\partial T}\right)_V = \left(\frac{\partial(A/T)}{\partial(1/T)}\right)_V \tag{7.13}$$

Carrying out the differentiation, and allowing for the possibility that $U(0) \neq 0$, we find we have recreated eqn 7.3

$$U - U(0) = kT^2\left(\frac{\partial \ln Q}{\partial T}\right)_V \tag{7.3}$$

Moreover, we can further our understanding by recasting this in some alternative forms

$$U - U(0) = -k\left(\frac{\partial \ln Q}{\partial(1/T)}\right)_V = -\left(\frac{\partial \ln Q}{\partial \beta}\right)_V \tag{7.14}$$

Equations 7.3 and 7.13 provide three of the basic relationships between the canonical partition function, Q, and the internal energy, U. Their relationship to the Helmholtz function, A, is stated explicitly in eqns 7.12 and 7.13. There is no surprise in this, just the satisfaction of knowing that the Massieu function can be used to recreate the internal energy function, U.

7.7 The equation of state and the pressure, *p*

The pressure, p, is related to the Helmholtz free energy through the first derivative with regard to volume at constant temperature.

$$p = -\left(\frac{\partial A}{\partial V}\right)_T \tag{7.15}$$

This simple relationship provides a direct route from the Massieu bridge to the equation of state

$$p = kT\left(\frac{\partial \ln Q}{\partial V}\right)_T \tag{7.16}$$

7.8 The heat capacity, C_V

The heat capacity, C_V, enjoys only a rather distant relationship with the Helmholtz free energy. It is much more directly related to the temperature derivative of eqn 7.3, in that

$$C_V = \left(\frac{\partial U}{\partial T}\right)_V \tag{7.17}$$

and rather less directly to the Massieu bridge. Nonetheless, one can write

$$C_V = \left\{\frac{\partial}{\partial T}\left[T^2\left(\frac{\partial(A/T)}{\partial T}\right)_V\right]\right\}_V \tag{7.18}$$

which, in turn, leads to an equation we saw earlier

$$C_V = \left\{\frac{\partial(U-U(0))}{\partial T}\right\}_V = \left\{\frac{\partial}{\partial T}\left[kT^2\left(\frac{\partial \ln Q}{\partial T}\right)_V\right]\right\}_V \tag{7.6}$$

$$= 2kT\left(\frac{\partial \ln Q}{\partial T}\right)_V + kT^2\left(\frac{\partial^2 \ln Q}{\partial T^2}\right)_V$$

7.9 The entropy, *S*

The entropy can be derived from the Massieu function quite simply

$$S = \left(\frac{\partial A}{\partial T}\right)_V \tag{7.19}$$

from which we obtain, as might be expected,

$$S = k\ln Q + kT\left(\frac{\partial \ln Q}{\mathrm{d}T}\right)_V \tag{7.7}$$

Once again, from the Massieu bridge we can recreate the correct expression for the entropy as derived previously in eqn 7.7 using other, less direct, means.

7.10 The enthalpy, *H*

At this point, the link between certain functions and the Massieu bridge can become a little distant. Nonetheless, since $H = U + pV$, we can use eqns 7.14 and 7.15 to give

$$H = T^2\left(\frac{\partial(A/T)}{\partial T}\right)_V - \left(\frac{\partial A}{\partial V}\right)_T \tag{7.20}$$

which, using eqns 7.3 and 7.16, leads to the final expression for the enthalpy

$$H - H(0) = kT^2\left(\frac{\partial \ln Q}{\partial T}\right)_V + kTV\left(\frac{\partial \ln Q}{\partial V}\right)_T \tag{7.21}$$

More directly, the result in eqn 7.21 can be obtained using the classical relationship $H = U + pV$

with $H(0)$, the enthalpy at the absolute zero, given by

$$H(0) = U(0) = A(0) \tag{7.22}$$

For a perfect gas, in which $pV = nkT$, eqn 7.21 can be combined with eqn 7.16 and recast in the form

$$H - H(0) = kT^2\left(\frac{\partial \ln Q}{\partial T}\right)_V + NkT \tag{7.23}$$

7.11 The Gibbs free energy, *G*

As with the enthalpy, we can express the Gibbs free energy in terms of the Helmholtz free energy, A, the pressure, p, and the volume, V, in the form $G = A + pV$, and, since we know p in terms of the Helmholtz free energy, we can write

$$G = A - V\left(\frac{\partial A}{\partial V}\right)_T \tag{7.24}$$

Using the Massieu bridge (eqn 7.12) we get

$$G - G(0) = -kT\ln Q + kTV\left(\frac{\partial \ln Q}{\partial V}\right)_T \tag{7.25}$$

where $G(0)$ is the Gibbs free energy at the absolute zero, another equality to add to eqn 7.22.

Equation 7.25 takes a particularly simple form for a perfect gas in which $pV = nkT$

$$G - G(0) = -kT\ln Q + nkT \tag{7.26}$$

with $G(0)$ identified with $U(0)$ and $A(0)$, as expected

$$U(0) = H(0) = A(0) = G(0) \tag{7.22a}$$

Now, for a gas consisting of N indistinguishable particles, we know that the canonical partition function, Q, is related to the molecular partition function, q, by

$$Q = \frac{q^N}{N!}$$

Stirling's approximation:
In exact form, the approximation for factorial x ($x!$) is
$$x! = (2\pi)^{1/2}x^{x+1/2}e^{-x}$$
For very large values of x, $x >> 1/2$ and the version quoted here,
$\ln x \approx x\ln x - x$ is perfectly acceptable.

so, using Stirling's approximation ($\ln N! \approx N\ln N - N$) we can write

$$-NkT\ln Q = -NkT\ln q + NkT\ln N - NkT$$

Furthermore, since the gas constant, R, the Boltzmann constant, k, and the Avogadro constant, L, are related

$$R = Lk \text{ and } nL = N$$

we have

$$nR = Nk$$

which leads to an overall expression for the Gibbs free energy

$$G - G(0) = -NkT\ln\left(\frac{q}{N}\right) = -nRT\ln\left(\frac{q}{nL}\right) \tag{7.27}$$

Equation 7.27 can be further transformed by defining two molar quantities, the quite familiar **molar Gibbs free energy**, G_m (with SI units J mol^{-1}), and the more novel **molar partition function,** q_m (with dimensions mol^{-1}), each obtained by dividing the extensive quantity by the number of moles, n.

$$G_m = \frac{G}{n} \text{ and } q_m = \frac{q}{n} \tag{7.28}$$

Doing this allows us to write another relationship for the Gibbs free energy, which is

$$G_m - G_m(0) = -RT \ln\left(\frac{q_m}{L}\right) \tag{7.29}$$

Equation 7.25, together with eqn 7.29 (derived at this juncture because it will be needed later in Chapter 15) add to the suite of toolkit equations we need in order to apply statistical thermodynamics to the real world which is embodied in classical thermodynamics. The most significant toolkit equations are summarised in Section 7.12.

7.12 A full set of toolkit equations

We recall below the results of Sections 7.5 to 7.11. These equations form enough of a comprehensive set of defining equations to be considered definitive, even though, as with all of classical thermodynamics, very many additional combinations and permutations of these equations lead to a wealth of ancillary relationships that link all the properties of matter. But these equations provide a basic working set of tools.

The full toolkit, then, is:

1. *Massieu bridge:* $-\dfrac{(A - A(0))}{T} = k \ln Q$

2. *Internal energy:* $U - U(0) = kT^2\left(\dfrac{\partial \ln Q}{\partial T}\right)_V$

3. *Equation of state:* $p = kT\left(\dfrac{\partial \ln Q}{\partial V}\right)_T$

4. *Heat capacity:* $C_V = 2kT\left(\dfrac{\partial \ln Q}{\partial T}\right)_V + kT^2\left(\dfrac{\partial^2 \ln Q}{\partial T^2}\right)_V$

5. *Entropy:* $S = k \ln Q + kT\left(\dfrac{\partial \ln Q}{\mathrm{d}T}\right)_V$

6. *Enthalpy:* $H - H(0) = kT^2\left(\dfrac{\partial \ln Q}{\partial T}\right)_V + kTV\left(\dfrac{\partial \ln Q}{\partial V}\right)_T$

7. *Gibbs function:* $G - G(0) = -kT \ln Q + kTV\left(\dfrac{\partial \ln Q}{\partial V}\right)_T$

7.13 Conclusions

Massieu bridge: $-A/T = k \ln Q$

In order to relate the partition function to classical thermodynamic quantities, two defining equations are needed. Convenient for this purpose are the **internal energy** and the **entropy** functions. Once these have been expressed in terms of the canonical partition function, Q, it is a simple matter to derive the **Massieu bridge**, which provides the most compact link possible to classical thermodynamics. The seven equations presented in Section 7.12, each of which can be derived from the Massieu function, form a coherent set of relationships with the aid of which the overwhelming majority of thermodynamic problems can be tackled.

The ease with which this is done can be a little misleading. Although we certainly know how to calculate thermodynamic properties once we have formulated the appropriate partition function for the problem in hand, knowing how to formulate the appropriate partition function is often far from simple; stating a problem in such a way as to derive the relevant partition function is rarely easy. This is a deceptively simple yet often intractable problem. It lies at the heart of much of current endeavour to use statistical thermodynamics to solve problems only slightly more complicated than the ones we shall tackle in subsequent chapters.

8 The ideal monatomic gas: the translational partition function

8.1 Introduction

The two-level system described in Chapter 6, while offering insight into uses of the partition function, has only limited applications in the real world. We need to add a measure of complexity to the systems we study, and the most straightforward way of doing this is to consider particles in which, in principle at least, there are very large numbers of accessible states for the system to sample. So next we shall consider an assembly of featureless particles constrained to move in a fixed volume. This system is one consisting of many, non-interacting, monatomic gas particles in ceaseless translational motion. The only energy that these particles can posses is translational kinetic energy.

8.2 The translational partition function, q_{trs}

In classical mechanics, all kinetic energies are allowed in a system of monatomic gas particles at a fixed volume V and temperature T. Quantum restrictions, however, place limits on the actual kinetic energies that are found. To determine the partition function for such a system, we need to establish values for the allowed kinetic energies. We do so by calling on a simple model, that of a particle constrained to move in a cubic box with dimensions ℓ_x, ℓ_y, and ℓ_z.

A particle in a one-dimensional box

The permitted energy levels, ε_x, for a particle of mass m that is constrained by infinite boundary potentials at $x = 0$ and $x = \ell_x$ to exist in a one-dimensional box of length ℓ_x are given by

$$\varepsilon_x = \frac{n_x^2 h^2}{8m\ell_x^2} \tag{8.1}$$

and similarly for the y- and z-directions. The translational quantum number, n_x, is a positive integer and the quantum numbers in the y- and z-directions are n_y and n_z, respectively.

The one-dimensional partition function, $q_{trs,x}$, is obtained by summing over all the accessible energy states. Thus

$$q_{trs,x} = \sum_{\text{all } n} e^{-\beta n_x^2 h^2 / 8m\ell_x^2} \tag{8.2}$$

an expression that is exact but which cannot be evaluated except by direct and tedious numerical summation.

Now, for all values of ℓ_x in any normal vessel, these energy levels are very densely packed and lie extremely close to each other. They form a **virtual continuum**, so it is quite proper to replace the summation in eqn 8.2 by an integration with the running variable n_x.

$$q_{\text{trs},x} = \int_0^\infty e^{-\beta n_x^2 h^2/8m\ell_x^2}\, dn_x \tag{8.3}$$

standard integral:

$$\int_0^\infty e^{-\alpha^2 u^2} du = \frac{\sqrt{\pi}}{2\alpha}$$

Furthermore, by changing the lower limit of integration from 1 to 0, as we have done above, negligible error is introduced and the resulting expression can then be evaluated in closed form using a standard integral. The result is

$$q_{\text{trs},x} = \left(\frac{2\pi m}{\beta}\right)^{\frac{1}{2}} \frac{\ell_x}{h} \tag{8.4}$$

Extension to three dimensions

Motion in the other two directions is independent of that in the x-direction, so we can immediately factorise the translational partition function and write

$$q_{\text{trs}} = q_{\text{trs},x} \times q_{\text{trs},y} \times q_{\text{trs},z} = \left(\frac{2\pi m}{\beta}\right)^{\frac{3}{2}} \frac{\ell_x \ell_y \ell_z}{h^3}$$

or, since the product of the three dimensions of the box yields the volume, V, of the box

It is worth noting that the final result is, in fact, quite independent of the shape of the "box" in which the particles are contained. It could be rectangular, spherical, or irregular in shape, although the derivation in the latter case is much less straightforward.

$$q_{\text{trs}} = \left(\frac{2\pi m}{h^2\beta}\right)^{\frac{3}{2}} V = \left(\frac{2\pi mkT}{h^2}\right)^{\frac{3}{2}} V \tag{8.5}$$

and, for the canonical partition function for N indistinguishable particles

$$Q_{\text{trs}} = \frac{q_{\text{trs}}^N}{N!} = \frac{1}{N!}\left[\left(\frac{2\pi m}{h^2\beta}\right)^{\frac{3}{2}} V\right]^N = \frac{1}{N!}\left[\left(\frac{2\pi mkT}{h^2}\right)^{\frac{3}{2}} V\right]^N \tag{8.6}$$

The continuum approximation: test of validity

It is worth testing the virtual continuum approximation used when the summation in eqn 8.2 was replaced by an integration. We do this by estimating the number of translational energy states accessible to a gas in a volume of reasonable laboratory dimensions, and we do so under the least favourable conditions, namely for the lightest available gas at the lowest temperature at which it can exist at atmospheric pressure.

The choice of gas is easy. Helium, the lightest monatomic gas, has a normal boiling temperature of 4.22 K. At this temperature, one mole of helium, assuming ideal gas behaviour, occupies 3.46×10^{-4} m^3, or some 350 cm^3, a reasonable size for a laboratory vessel. By calculating q_{trs} for one mole of helium at its normal boiling point we will establish how many translational energy states are accessible at this very low temperature and conclude whether or not the virtual continuum approximation is valid.

In addition, by calculating the ratio q_{trs}/L, we will be able determine how many states are accessible to each helium atom and hence decide if the criterion for the valid use of Boltzmann statistics ($q/L \gg 1$) is in place.

First, it is helpful to gather together all the constants in eqn 8.5 to yield

$$q_{trs} = \left(\frac{2\pi k}{h^2 L}\right)^{\frac{3}{2}} (MT)^{\frac{3}{2}} V$$

which, in SI units, becomes

$$q_{trs} = \left(5.942 \times 10^{30} / \mathrm{mol}^{\frac{5}{2}} \mathrm{m}^{-3} \mathrm{kg}^{-\frac{3}{2}} \mathrm{K}^{-\frac{5}{2}}\right) (MT)^{\frac{3}{2}} V \qquad (8.7)$$

where the rather bizarre units of the constant arise from the dimensions of the fundamental constants involved, leading to a translational partition function that is dimensionless.

Inserting into eqn 8.7 the value 4.003×10^{-3} kg mole^{-1} for M_{He}, 4.22 K for T_b, and 3.46×10^{-4} m^3 mole^{-1} for V, the result obtained is

$$q_{trs} = 4.5 \times 10^{24}$$

For a heavier molecule, CO_2 for example, $q_{trs} \approx 10^{30}$ at room temperature in a vessel of laboratory dimensions.

The very large number of states accessible to gaseous helium atoms at the boiling point gives credence to the virtual continuum approximation.

States available to each particle

Although q_{trs} above is indeed large, the ratio q_{trs}/L is not unduly so. For helium at its boiling point, q_{trs}/L has the value 7.5. While this value does exceed unity, it does not do so by a very wide margin, though it is still realistic to postulate that, by and large, each atom exists in its own, singly occupied, translational state.

But when q/L drops below unity, as happens in *liquid* helium at 4.22 K (where the molar volume is smaller by a factor of about 10) it becomes necessary to abandon Boltzmann statistics and to invoke more specialised (Bose–Einstein and Fermi–Dirac) **quantum statistics**. The problems thus uncovered, and their resolution, lie well outside the scope of this book. One should, nonetheless, be encouraged by the realisation that, while these problems do exist, methods also exist to solve them.

8.3 The ideal monatomic gas: thermodynamic functions

Since all classical thermodynamic functions are related to the logarithm of the canonical partition function, we start by taking the logarithm of eqn 8.6.

$$\ln Q_{trs} = \tfrac{3}{2} N \ln(2\pi m) + \tfrac{3}{2} N \ln T + N \ln V - 3N \ln h - \ln N! \qquad (8.8)$$

Only the second and third terms in eqn 8.8 have functional dependence on temperature or volume; so the derivatives of $\ln Q$ turn out to be very simple.

Toolkit functions

$$U = kT^2 \left(\frac{\partial \ln Q}{\partial T}\right)_V$$

$$p = kT \left(\frac{\partial \ln Q}{\partial V}\right)_T$$

$$C_V = 2kT \left(\frac{\partial \ln Q}{\partial T}\right)_V + kT^2 \left(\frac{\partial^2 \ln Q}{\partial T^2}\right)_V$$

$$S = k \ln Q + kT \left(\frac{\partial \ln Q}{\partial T}\right)_V$$

$$\left(\frac{\partial \ln Q_{trs}}{\partial T}\right)_V = \frac{3N}{2T} \quad \text{and} \quad \left(\frac{\partial^2 \ln Q_{trs}}{\partial T^2}\right)_V = -\frac{3N}{2T^2} \qquad (8.9a)$$

and

$$\left(\frac{\partial \ln Q_{trs}}{\partial V}\right)_T = \frac{N}{V} \qquad (8.9b)$$

which can be used in the toolkit functions shown earlier.

The translational energy

For a monatomic gas, the translational kinetic energy, E_{trs}, is the only form of energy that the monatomic particles posses, so we can equate it directly to the internal energy, U, and, substituting the value of the derivative (eqn 8.9a)

$$E_{trs} = U = kT^2\left(\frac{\partial \ln Q_{trs}}{\partial T}\right)_V = \tfrac{3}{2}NkT$$

and, for one mole ($N = L$, $Lk = R$)

Equipartition of energy: From the kinetic theory of gases we can conclude that each degree of translational freedom (there are three, in mutually perpendicular directions) carries an equal amount of energy (hence *equipartition*), namely $\frac{1}{2}kT$. So, in three dimensions $U_m = \frac{3}{2}RT$

$$U_m = \tfrac{3}{2}RT \qquad (8.10)$$

as might be expected from the equipartition of energy in an ideal monatomic gas.

The equation of state

The equation of state can be derived using the toolkit expression for the pressure p and substituting the value of the derivative (eqn 8.9b)

$$p = kT\left(\frac{\partial \ln Q_{trs}}{\partial V}\right)_T = \frac{NkT}{V}$$

and, for one mole ($N = L$, $Lk = R$)

The ideal gas equation: $pV_m = RT$

$$p = \frac{RT}{V_m} \quad \text{or} \quad pV_m = RT \qquad (8.11)$$

again exactly as could be predicted from the properties of the ideal monatomic gas.

The heat capacity

Using the toolkit expression for the heat capacity at constant volume, C_V, and, substituting values of both derivatives (eqn 8.9a)

$$C_V = 2kT\left(\frac{\partial \ln Q_{trs}}{\partial T}\right)_V + kT^2\left(\frac{\partial^2 \ln Q_{trs}}{\partial T^2}\right)_V = 3Nk - \tfrac{3}{2}Nk$$

and, for one mole ($N = L$, $Lk = R$)

The ideal gas heat capacity: $C_{V,m} = \frac{3}{2}R$

$$C_{V,m} = \tfrac{3}{2}R \qquad (8.12)$$

In all three of the expressions above (eqns 8.10 – 8.12), the outcome of using the translational **canonical partition function** is entirely in agreement with the results already well known from the **kinetic theory of gases** and **classical thermodynamics**. In the next section, devoted to the translational entropy, we shall discover some new features and not merely recover information already established by classical thermodynamics

8.4 The entropy of the ideal monatomic gas

When it comes to using statistical thermodynamics to determine the translational entropy of an ideal monatomic gas, it is immediately clear that all of the simplifying factors that result from taking partial derivatives of $\ln Q_{trs}$ no longer hold.

$$S = k \ln Q_{trs} + kT\left(\frac{\partial \ln Q_{trs}}{\partial T}\right)_V \qquad (8.13)$$

The statistical mechanical expression for the entropy (eqn 8.13) contains a direct term in $\ln Q_{trs}$, so all of the variables in eqn 8.8 are going to appear in the final expression for the entropy, S.

We start with the first term in eqn 8.13 and revert to the formulation involving the molecular partition function, q_{trs}.

$$k \ln Q_{trs} = k \ln \frac{1}{N!} q^N$$

$$= k\left(\ln \frac{1}{N!} + \ln q^N\right) = k(N \ln q - \ln N!)$$

Next, using Stirling's approximation ($\ln N! \approx N \ln N - N$) we can write

$$k \ln Q_{trs} = k(N \ln q - N \ln N + N) = Nk\left(1 + \ln \frac{q}{N}\right)$$

and, substituting this into eqn 8.13 together with eqn 8.9a, we get

$$S = Nk\left(1 + \ln \frac{q}{N}\right) + \frac{3}{2}Nk = Nk\left(\frac{5}{2} + \ln \frac{q}{N}\right)$$

Finally, using the expression for q_{trs} (eqn 8.5), the resulting entropy is

$$S = Nk\left\{\frac{5}{2} + \ln\left[\left(\frac{2\pi mkT}{h^2}\right)^{\frac{3}{2}} \frac{V}{N}\right]\right\} \qquad (8.14)$$

For one mole of an ideal gas ($N = L$, $Lk = R$, $Lm = M$ and $V = RT/p$)

$$S_m = R\left\{\frac{5}{2} + \ln\left[\left(\frac{2\pi MkT}{h^2 L}\right)^{\frac{3}{2}} \frac{RT}{Lp}\right]\right\}$$

When all the experimental variables (M, T, and p) are gathered together, the resulting expression is

$$S_m = R\left[\ln\left(\frac{M^{\frac{3}{2}}T^{\frac{5}{2}}}{p}\right)\right] + R\left\{\ln\left[\left(\frac{2\pi}{Lh^2}\right)^{\frac{3}{2}}(\mathrm{e}k)^{\frac{5}{2}}\right]\right\} \quad (8.15)$$

THE SACKUR–TETRODE EQUATION

8.5 Using the Sackur–Tetrode equation

Let us examine the Sackur–Tetrode equation as it appears in eqn 8.15. The first term contains all the experimental variables: the molar mass of the test gas, its temperature, and its pressure. These are the parameters that will vary from one situation to another.

The second term is simply a constellation of fundamental constants, independent of the choice of either the test substance or the given experimental conditions. Its value, in SI units, is 172.29 J K^{-1} $mole^{-1}$, or 20.723 *R*. It is simply a number whose value depends only on the unit system being used and, as such, adds nothing to our understanding of factors that influence the entropy of ideal monatomic gases. Consequently, it is more illuminating to write the Sackur–Tetrode equation without being explicit about the composition of this constant, while retaining the confidence that its composition and, indeed, its value is well known. This is done in eqn 8.16 below.

$$S_m/R = \ln\left[\left(M/\text{kg mol}^{-1}\right)^{\frac{3}{2}}(T/\text{K})^{\frac{5}{2}}(p/\text{Pa})^{-1}\right] + constant_1 \quad (8.16)$$

where $constant_1$ takes the value 20.723.

Table 8.1 Calculated and calorimetric entropies

	Argon	Krypton
T_b/K	87.4	120.2
S_{calc}/R	15.542	17.451
S_{calor}/R	15.60	17.43

An alternative expression, simpler to use, but less transparent in terms of the origin of its numerical constant, can be written as

$$S_m/R = \ln\left[M_r^{\frac{3}{2}}(T/\text{K})^{\frac{5}{2}}(p/p^{\ominus})^{-1}\right] + constant_2$$

where $constant_2$ takes the value −1.165 when the standard state pressure is chosen as $p^{\ominus} = 10^5$ Pa.

In Table 8.1, the entropy of argon and krypton at their respective boiling temperatures, calculated using eqn 8.16, are compared with experimental, calorimetric values. The agreement is excellent. Indeed, the calculated values are considered to be the more reliable ones, as is indicated by the number of significant figures shown.

Apart from making it possible to calculate absolute entropies, what else does the Sackur–Tetrode equation show? It is possible to recast eqn 8.16 in terms of dependence on volume rather than pressure. Using the ideal gas expression for $\ln p^{-1}$ and expanding the logarithmic terms, gives

$\frac{1}{p} = \frac{V}{RT}$ so

$\ln p^{-1} = \ln V - \ln R - \ln T$

$$S_m = R\ln V + \tfrac{3}{2}R\ln T + \tfrac{3}{2}R\ln M + constant_3 \quad (8.17)$$

$constant_3/R = 20.723 - \ln R = 18.605$

with the constant now taking the value 18.605 *R* when the variables are expressed in SI units (V/m^3, *T*/K, and *M*/kg mol^{-1}).

Examining eqn 8.17 term by term, as is done in Scheme 8.1, we see that the first two terms predict correctly the well-known volume and temperature dependence of the entropy of an ideal gas. The last two terms, however, could not be foreseen from classical thermodynamics. The mass dependence could perhaps have been inferred empirically, but without providing the clear physical insight given by eqn 8.17. This quantitative insight is a major new success for the analysis provided by statistical thermodynamics.

The conglomeration of constants (the last term), on the other hand, is quite novel.

$$S_m \;=\; R\ln V \;+\; \tfrac{3}{2}R\ln T \;+\; \tfrac{3}{2}R\ln M \;+\; 18.605\,R$$

$$\Uparrow \qquad\qquad \Uparrow$$

Volume dependence   *Temperature dependence*   *Could not be foreseen from classical thermodynamics*

$$\Delta S_T = R\ln\frac{V_2}{V_1} \qquad \Delta S_V = \tfrac{3}{2}R\ln\frac{T_2}{T_1}$$

$$= C_V \ln\frac{T_2}{T_1}$$

Known from classical thermodynamics

Scheme 8.1   Components of the Sackur–Tetrode equation

Analysis of the Sackur–Tetrode equation, as shown above, gives insight into the way in which details of the translational partition function affect classical thermodynamic functions.

Thus, an increase in **volume** or in molecular **mass**, at **constant temperature**, decreases the spacing between adjacent energy states (see e.g. the one-dimensional case, eqn 8.1). The **density of states** (the number of states in unit energy interval) increases, making more states accessible at fixed temperature (eqn 8.5). This affects the entropy (because there are more ways in which the total energy can be distributed over the system), but not the internal energy, the heat capacity, or the pressure, which depend only on the temperature itself, or on the volume derivative of $\ln Q_{trs}$ (eqn 8.9b).

Density of states function $D(\varepsilon) = \dfrac{dN(\varepsilon)}{d\varepsilon}$

An increase in temperature, at **constant volume** and **mass**, on the other hand, does not affect the density of states; more states become accessible because the **thermal energy**, kT, increases. Temperature derivatives of $\ln Q_{trs}$ have an inverse dependence on temperature; these result in U and p becoming directly proportional to absolute temperature while leaving C_v independent of it. The entropy increases because more states are accessible.

These conclusions are summarised in Table 8.2.

Table 8.2   Changes in conditions

	Effect of increase in T, V, or M, on					
	$D(\varepsilon)$	Q	U	p	C_v	S
T	nil	↑	↑	↑	nil	↑
V	↑	↑	nil	nil	nil	↑
M	↑	↑	nil	nil	nil	↑

8.6 Conclusions

Statistical thermodynamics allows us to calculate theoretical values of entropies for monatomic gases to compare with (and often to replace) experiment. It also helps to bring together two major threads in our understanding of thermodynamics: the statistical and thermomechanical views of entropy.

9 The ideal diatomic gas: internal degrees of freedom

9.1 Introduction

The energy of **translation, rotation, vibration** and **electronic excitation**.

In the last chapter we gained an insight into the behaviour of ideal monatomic gas particles whose only energy was the kinetic energy of translational motion. Polyatomic species, in contrast, can store energy in a variety of other ways, most obviously as

- translational motion
- rotational motion
- vibrational motion

as well as

- electronic excitation

although this last is not exclusively in the provenance of polyatomic species, and could, quite properly, have been added to the properties of monatomic species in Chapter 8. However, thermal electronic excitation in monatomic species is rarely significant, so it makes good sense to defer this discussion until the end of this chapter.

Given that molecules containing more than one atom have these additional ways of storing energy, each mode with its own manifold of energy states, how do we set about coping with this added complexity?

9.2 Internal modes: separability of energies

Energy modes are **separable**.

The fact that molecule *i* is in translational energy state *s* has no direct effect on the internal energy states (rotation, vibration, electronic) that it occupies.

We start by invoking the assumption that molecular energy modes are **separable** in the sense that we can treat each mode as being essentially **independent** of all the other modes. This is strictly true as regards translational energy, which is quite independent of all the other modes. And it is essentially true for other modes if certain reasonable assumptions are made. Vibrational modes, for example, are independent of rotational modes under the **rigid rotor** assumption, and are also independent of electronic modes under the **Born–Oppenheimer** approximation.

A molecule that is moving at high speed is not, as a consequence, forced to rotate very fast or to vibrate rapidly. These modes are independent in the sense that an isolated molecule which has an excess of any one energy mode cannot, in general, divest itself of this surplus except at collision with another molecule. The number of collisions needed to equilibrate modes varies from a few (of the order of ten or so) for rotation, to many (of the order of hundreds) for vibration. So internal modes behave in an essentially independent manner in assemblies of molecules. The different internal modes are separable in individual molecules and each enjoys its own Boltzmann energy distribution, unperturbed by the presence of other modes. So modes are separable in individual molecules and, on average, are equally so in large assemblies of molecules.

This separability of energy modes allows us to write for the total energy of a molecule j

$$\varepsilon_{\text{tot}}^{j} = \varepsilon_{\text{trs}}^{j} + \varepsilon_{\text{rot}}^{j} + \varepsilon_{\text{vib}}^{j} + \varepsilon_{\text{el}}^{j} \tag{9.1}$$

9.3 Weak coupling: factorising the energy modes

At the heart of the assumption of separability is the concept of **weak coupling**, which, while admitting some energy interchange in order to establish and maintain thermal equilibrium (i.e. not zero coupling), allows us to assess each energy mode as if it were the only form of energy present in the molecule.

Thermal equilibrium

The outcome of the weak coupling assumption is to enable us to formulate the molecular partition function separately for each energy mode or degree of freedom and, only later, to decide how these individual partition functions should be combined together to form the overall molecular partition function.

Separability of translational modes in three dimensions

Just how this combination works is most easily seen in a hypothetical but very simple case. Imagine an assembly of N particles that can store energy in just two weakly coupled modes. For convenience, we can call these modes the *alpha* mode (α) and the *omega* (ω) mode. Each of these modes has its own manifold of energy states and associated quantum numbers. A given particle, therefore, could have alpha-mode energy associated with quantum number k, say, and omega-mode energy associated with quantum number r. The total energy of this particle would be

$$\varepsilon_{\text{tot}} = \varepsilon_{\alpha k} + \varepsilon_{\omega r} \tag{9.2}$$

where the subscripts αk and ωr serve to identify both the energy manifold in question and the quantum state occupied within that manifold.

For the overall partition function, q_{tot}, we would write, in brief notation

$$q_{\text{tot}} = \sum_{\textit{all states}} e^{-\beta(\varepsilon_{\alpha i} + \varepsilon_{\omega j})}$$

and, expanding the summation by writing the first few terms we would get

$$\begin{aligned}
q_{\text{tot}} = {} & e^{-\beta(\varepsilon_{\alpha 0}+\varepsilon_{\omega 0})} + e^{-\beta(\varepsilon_{\alpha 0}+\varepsilon_{\omega 1})} + e^{-\beta(\varepsilon_{\alpha 0}+\varepsilon_{\omega 2})} + e^{-\beta(\varepsilon_{\alpha 0}+\varepsilon_{\omega 3})} + \ldots \\
& + e^{-\beta(\varepsilon_{\alpha 1}+\varepsilon_{\omega 0})} + e^{-\beta(\varepsilon_{\alpha 1}+\varepsilon_{\omega 1})} + e^{-\beta(\varepsilon_{\alpha 1}+\varepsilon_{\omega 2})} + e^{-\beta(\varepsilon_{\alpha 1}+\varepsilon_{\omega 3})} + \ldots \\
& + e^{-\beta(\varepsilon_{\alpha 2}+\varepsilon_{\omega 0})} + e^{-\beta(\varepsilon_{\alpha 2}+\varepsilon_{\omega 1})} + e^{-\beta(\varepsilon_{\alpha 2}+\varepsilon_{\omega 2})} + e^{-\beta(\varepsilon_{\alpha 2}+\varepsilon_{\omega 3})} + \ldots \\
& + e^{-\beta(\varepsilon_{\alpha 3}+\varepsilon_{\omega 0})} + e^{-\beta(\varepsilon_{\alpha 3}+\varepsilon_{\omega 1})} + e^{-\beta(\varepsilon_{\alpha 3}+\varepsilon_{\omega 2})} + e^{-\beta(\varepsilon_{\alpha 3}+\varepsilon_{\omega 3})} + \ldots \\
& + \ldots
\end{aligned}$$

Now, the properties of exponentials allow us to write $e^{(a+b)}$ as $e^a.e^b$, so each of the terms in the expansion above can be written as the product of two exponentials, one for mode α, the other for mode ω. Doing this yields

$$\begin{aligned} q_{tot} = \; & e^{-\beta\varepsilon_{\alpha 0}}.e^{-\beta\varepsilon_{\omega 0}} + e^{-\beta\varepsilon_{\alpha 0}}.e^{-\beta\varepsilon_{\omega 1}} + e^{-\beta\varepsilon_{\alpha 0}}.e^{-\beta\varepsilon_{\omega 2}} + \ldots \\ & + e^{-\beta\varepsilon_{\alpha 1}}.e^{-\beta\varepsilon_{\omega 0}} + e^{-\beta\varepsilon_{\alpha 1}}.e^{-\beta\varepsilon_{\omega 1}} + e^{-\beta\varepsilon_{\alpha 1}}.e^{-\beta\varepsilon_{\omega 2}} + \ldots \\ & + e^{-\beta\varepsilon_{\alpha 2}}.e^{-\beta\varepsilon_{\omega 0}} + e^{-\beta\varepsilon_{\alpha 2}}.e^{-\beta\varepsilon_{\omega 1}} + e^{-\beta\varepsilon_{\alpha 2}}.e^{-\beta\varepsilon_{\omega 2}} + \ldots \\ & + \ldots \end{aligned}$$

Each term in every row of the expansion shown above has a common factor, $e^{-\beta\varepsilon_{\alpha 0}}$ in the first row, $e^{-\beta\varepsilon_{\alpha 1}}$ in the second, $e^{-\beta\varepsilon_{\alpha 2}}$ in the third, and so on. Extracting these factors, row by row, gives

$$\begin{aligned} q_{tot} = \; & e^{-\beta\varepsilon_{\alpha 0}}(e^{-\beta\varepsilon_{\omega 0}} + e^{-\beta\varepsilon_{\omega 1}} + e^{-\beta\varepsilon_{\omega 2}} + e^{-\beta\varepsilon_{\omega 3}} + \ldots) \\ & + e^{-\beta\varepsilon_{\alpha 1}}(e^{-\beta\varepsilon_{\omega 0}} + e^{-\beta\varepsilon_{\omega 1}} + e^{-\beta\varepsilon_{\omega 2}} + e^{-\beta\varepsilon_{\omega 3}} + \ldots) \\ & + e^{-\beta\varepsilon_{\alpha 2}}(e^{-\beta\varepsilon_{\omega 0}} + e^{-\beta\varepsilon_{\omega 1}} + e^{-\beta\varepsilon_{\omega 2}} + e^{-\beta\varepsilon_{\omega 3}} + \ldots) \\ & + e^{-\beta\varepsilon_{\alpha 3}}(e^{-\beta\varepsilon_{\omega 0}} + e^{-\beta\varepsilon_{\omega 1}} + e^{-\beta\varepsilon_{\omega 2}} + e^{-\beta\varepsilon_{\omega 3}} + \ldots) \\ & + \ldots \end{aligned}$$

The terms in parentheses in each row are identical and form the summation

$$\sum_{\textit{all } \omega \textit{ states}} e^{-\beta\varepsilon_{\omega j}}$$

which can, in turn, be factored out from each row, giving

$$\begin{aligned} q_{tot} &= \left(e^{-\beta\varepsilon_{\alpha 0}} + e^{-\beta\varepsilon_{\alpha 1}} + e^{-\beta\varepsilon_{\alpha 2}} + e^{-\beta\varepsilon_{\alpha 3}} + \ldots\right)\left(\sum_{\textit{all } \omega \textit{ states}} e^{-\beta\varepsilon_{\omega j}}\right) \\ &= \sum_{\textit{all } \alpha \textit{ states}} e^{-\beta\varepsilon_{\alpha j}} \times \sum_{\textit{all } \omega \textit{ states}} e^{-\beta\varepsilon_{\omega j}} \end{aligned}$$

where the sum in the first pair of parentheses is recognisably the sum over all the α states or, in other words, the partition function for the α energy modes.

So we conclude that, if energy modes are separable, as assumed in eqn 9.2, then we can factorise the partition function and write

$$q_{tot} = q_\alpha \times q_\omega \tag{9.3}$$

The extension of this reasoning to three or more weakly coupled, separable, modes is a trivial and self-evident exercise.

9.4 Factorising translational energy modes

We pre-empted the result in eqn 9.3 implicitly in Chapter 8 where, in deriving the three-dimensional translational partition function (eqn 8.5), we deriving the three-dimensional translational partition function (eqn 8.5), we assumed that the three components of translational motion were completely independent of each other. We can now develop this idea further and establish procedures for factorised partition functions that can be followed in all cases where molecular energies are separable.

By invoking separability, we imply (cf. eqn 9.1) that, for molecule j

$$\varepsilon^j_{\text{trs,tot}} = \varepsilon^j_{\text{trs},x} + \varepsilon^j_{\text{trs},y} + \varepsilon^j_{\text{trs},z} \tag{9.4}$$

which allows us to write

$$q_{\text{trs}} = \sum_{\textit{all states}} e^{-\beta\varepsilon_{\text{trs,tot}}} = \sum_{\textit{all states}} e^{-\beta(\varepsilon_{\text{trs},x}+\varepsilon_{\text{trs},y}+\varepsilon_{\text{trs},z})}$$

$$\therefore\ q_{\text{trs}} = \sum_{\textit{all x states}} e^{-\beta\varepsilon_{\text{trs},x}} \times \sum_{\textit{all y states}} e^{-\beta\varepsilon_{\text{trs},y}} \times \sum_{\textit{all z states}} e^{-\beta\varepsilon_{\text{trs},z}}$$

so

$$q_{\text{trs}} = q_{\text{trs},x} \times q_{\text{trs},y} \times q_{\text{trs},z} \tag{9.5}$$

Thus the assumption of separable energy modes (weakly coupled, by inference) allows us to **factorise** the molecular partition function into individual components.

Factorisation of the partition function.

9.5 Factorising internal energy modes

It is no surprise then, that, invoking the separability expressed in eqn 9.1, we can write an expression for the overall molecular partition function as follows

$$q_{\text{tot}} = q_{\text{trs}} \cdot q_{\text{rot}} \cdot q_{\text{vib}} \cdot q_{\text{el}} \tag{9.6}$$

Using identical arguments, the canonical partition function can also be factorised in this manner,

$$Q_{\text{tot}} = Q_{\text{trs}} \cdot Q_{\text{rot}} \cdot Q_{\text{vib}} \cdot Q_{\text{el}} \tag{9.7}$$

But how do we obtain the canonical from the molecular partition function, Q_{tot} from q_{tot}? How does indistinguishability exert its influence?

The answer to these questions is crucial, if we are to take advantage of the simplification afforded by factorisation, but it is also logical and simple. We need to answer the question: when are particles distinguishable as regards creating distinct configurations, and when are they indistinguishable?

Indistinguishability factorisation.

We have already argued that **localised** particles (with unique addresses) are always **distinguishable**. Equally, particles that are **not localised** but move freely are **indistinguishable**, since swapping of translational energy states between such particles does not create distinct new configurations.

Distinguishable energy modes $Q = q^N$

Indistinguishable modes $Q = \dfrac{q^N}{N!}$

But localisation in space is not the only way of conferring an address and with it a distinguishable identity. Localisation within a molecule can also confer distinguishability. When molecules i and j, each in distinct rotational and vibrational states, swap these internal states with each other, a new configuration is created, and *both* configurations have to be counted into the final **sum of states** for the whole system. By being identified specifically with individual molecules, the internal states are recognised as being intrinsically **distinguishable**.

Only translational energies can create indistinguishable configurations; all other energies cannot, because they are **localised** on the particles to which they belong.

Particles in solids are a special case: since they lack free translational motion, interchanges in the captive energy they posses always leads to new configurations.

But if i and j swap their translational energies instead, a new configuration is not created; translational states are thus recognised as being intrinsically **indistinguishable**, and they too are treated as such. It is only the interchange of the kinetic energy of free translation that creates no new configurations. In a certain sense, once the element of uncertainty we call indistinguishability has been invoked, all other energy states are tied to intrinsically distinguishable particles.

It is perhaps most helpful to consider indistinguishability to be a property not so much of the particles themselves, but more of the configurations to which they can or cannot give rise.

9.6 The canonical partition function, Q

Applying this reasoning to eqn 9.7, we get

$$Q_{\text{tot}} = \frac{(q_{\text{trs}})^N}{N!}(q_{\text{rot}})^N (q_{\text{vib}})^N (q_{\text{el}})^N$$

whence

$$Q_{\text{tot}} = \frac{1}{N!}\left(q_{\text{trs}} \cdot q_{\text{rot}} \cdot q_{\text{vib}} \cdot q_{\text{el}}\right)^N \tag{9.8}$$

Weak coupling

This conclusion depends on the assumption of weak coupling. If particles enjoy **strong coupling** (e.g. in liquids and solutions), the argument becomes very complicated. Such problems are currently of great interest in modern statistical thermodynamics, and the answers to the problems posed are at the forefront of research in this subject.

9.7 Conclusions

Now that we know how to divide up the total partition function, we must turn to the problem of evaluating each of the factors that are created by this division.

Eqn 8.5:

$$q_{\text{trs}} = \left(\frac{2\pi m}{h^2 \beta}\right)^{\frac{3}{2}} V = \left(\frac{2\,\pi mkT}{h^2}\right)^{\frac{3}{2}} V$$

We started with external (translational) modes of motion. These pose no new problems when it comes to considering particles with structure that have internal modes. Equation 8.5, shown opposite, can be applied to molecules containing more than one atom just as validly as they were applied to monatomic species in Chapter 8.

Working expressions for the individual partition functions for internal motion are developed fully in the chapters that follow.

10 The ideal diatomic gas: the rotational partition function

10.1 Introduction

Species that have more than one atom can store energy in the form of rotational motion. As with all other forms of energy, rotational motion is restricted to certain values of the available energy, so molecules have accessible to them a manifold of **quantised rotational energy levels.**

As we shall see in Section 10.4, energy separations in the rotational manifold, while many times greater than the separations in translational energy, are not in themselves comparable in size to the thermal energy kT, except for very light molecules, or at very low temperatures, or both.

If we compare the **characteristic rotational temperature** (defined in Section 10.2) with ambient temperature, say, we can see that even in hydrogen, which has by far the largest rotational energy spacings of all molecules, the characteristic rotational temperature, $\theta_r = 88$ K, is not very large. Consequently, we can expect excited rotational energy states to be populated to an appreciable extent, though not nearly as much as translational states, for which energy level spacings are very much smaller indeed.

Comparing θ_r with T is equivalent to comparing $\Delta\varepsilon_r$ with kT.

10.2 The rigid rotor

We start by making the **rigid rotor** assumption, for which we can write successive rotational energy levels, ε_J, in terms of the **rotational quantum number,** J.

$$\varepsilon_J = J(J+1)\frac{h^2}{8\pi^2 I} = hcBJ(J+1) \text{ with } I = \mu r^2 \text{ and } B = \frac{h}{8\pi^2 Ic} \quad (10.1)$$

For a diatomic molecule, with atoms of mass m_1 and m_2

$$\mu = \frac{m_1 m_2}{m_1 + m_2}$$

where I is the **moment of inertia** of the molecule, μ the **reduced mass,** and B the **rotational constant**.

An alternative, and much simpler, expression results from using the **characteristic rotational temperature,** θ_r, defined opposite. This gives

$$\varepsilon_J = J(J+1)k\theta_r \quad (10.2)$$

Characteristic rotational temperature

$$\theta_r = \frac{h^2}{8\pi^2 Ik} = \frac{hcB}{k}$$

Note that θ_r does not correspond to any actual energy gap in the rotational manifold; it is more of a characteristic factor, which then is modified by the rotational quantum number, $J(J+1)$. Even the lowest gap ($J = 0 \rightarrow J = 1$) corresponds to an energy increment of $2k\theta_r$, and the next to $4k\theta_r$, and so on.

As it rotates, a molecule undergoes **centrifugal stretching.** This increases the moment of inertia, causing a small adjustment to the spacing of the rotational levels. This adjustment can be quantified as $\varepsilon_j = J(J+1)k\theta_r - D_J J^2(J+1)^2$ with the centrifugal constant, D_J, a function of each individual molecule. Such a correction affects the rotational partition function by less than 0.1% at temperatures below 500 K and can usually be ignored.

Rotational energy levels are **degenerate** and each level has a degeneracy $g_J = (2J + 1)$. So

$$q_{\text{rot}} = \sum g_J \, e^{-\varepsilon_J/kT} = \sum (2J+1)\, e^{-J(J+1)\theta_r/T} \tag{10.3}$$

Unfortunately, the series in the summation of eqn 10.3 does not sum to a closed analytical function but, if no atoms in the molecule are too light (i.e. if the moment of inertia, I, is not too small) and if the temperature is not too low (i.e. not too close to 0 K), allowing appreciable numbers of rotational states to be occupied, the rotational energy levels lie sufficiently close to each other to allow us, once again, to use the virtual continuum approximation and to replace the summation by an integration.

$$q_{\text{rot}} = \int_0^{\infty} (2J+1)\, e^{-J(J+1)\theta_r/T}\, dJ \tag{10.4}$$

$$\int_0^{\infty} e^{-ax}\, dx = \frac{1}{a}$$

Noting that $(2J+1)\,dJ = d(J(J+1))$, we can express the integral in eqn 10.4 in the form $\int e^{-ax}\,dx$, with $a = \theta_r/T$ and $x = J(J+1)$, and then, using the standard solution for this integral, obtain the result given in eqn 10.5.

$$q_{\text{rot}} = \frac{T}{\theta_r} \tag{10.5}$$

This deceptively simple result requires more careful thought in many cases, as we shall see shortly.

10.3 The continuum approximation: test of validity

$$q_{\text{rot}} = \frac{T}{\theta_r} = \frac{8\pi^2 IkT}{h^2}$$

As for translational energy in Chapter 8, it is worth testing the virtual continuum approximation used when changing eqn 10.3 from summation to integration. The outcome here, however, is not as clear-cut as it was for translation, although the criterion we seek to apply – that many (rotational) states should be thermally accessible – is the same in both cases.

Virtual continuum approximation

$T/\theta_r \geq 100$: no appreciable error

$T/\theta_r = 10$: about 0.5% error

The approximation is in error:
- for light molecules, and
- at low temperatures

We do this by estimating the number of rotational energy states accessible to a gas at different temperatures. We can start with the uncontentious view that if $T/\theta_r \geq 100$, (i.e. $q_{\text{rot}} \geq 100$), then the virtual continuum approximation is fully justified. Even at $T/\theta_r = 10$, the difference between summation and integration is less than 0.5%. At $T/\theta_r = 1$, integration gives an answer some 15% too large; at $T/\theta_r = 2$, the overestimate drops to 2%. In general, the approximation fails if θ_r is too big (i.e. the moment of inertia, I, is too small, as with light molecules) or if the temperature is too low.

The moment of inertia is small in diatomic gases that contain hydrogen. If there is a hydrogen atom in a diatomic molecule, then $T/\theta_r \sim 10$ at the normal boiling temperature. Usually, the continuum approximation can be used if the temperature does not drop too far below room temperature.

For diatomics that do not contain hydrogen, $T/\theta_r > 100$ for all temperatures at which the diatomic species remains gaseous. So problems will only arise at very low temperatures in molecules containing hydrogen.

10.4 Accessible states and symmetry

Equation 10.5 works with complete reliability (subject to reservations with light atoms, as outlined above) for all **heteronuclear** diatomic molecules. For **homonuclear** diatomics, however, a new consideration arises.

Equation 10.5 overcounts the rotational states available to homonuclear diatomic molecules. It does so by a factor of two. The simplest way to visualise the reason for this is to recognise that when a symmetrical linear molecule rotates through 180°, it produces a configuration that is **indistinguishable** from the one from which it started. Heteronuclear diatomic molecules do not have such a symmetry property and must rotate through 360° before reaching an indistinguishable state.

For homonuclear diatomic molecules (e.g. O_2, N_2, Cl_2), q_{rot} calculated using eqn 10.5 is **too large** by a factor of 2.

For symmetrical molecules, which includes all homonuclear diatomics as well as other symmetrical linear molecules (such as carbon dioxide and ethyne) it is important to recognise the existence of this **rotational indistinguishability**. For such molecules, the value of q_{rot} calculated using eqn 10.5 must be divided by a factor of 2 to yield the correct result. Better still, we can include all molecules within a single expression by introducing a **symmetry factor**, σ, which takes the value 2 for homonuclear diatomics and 1 for heteronuclear diatomics. Thus

$$q_{rot} = \frac{T}{\sigma \theta_r} \qquad (10.6)$$

Symmetry factor

σ=1 for heteronuclear diatomics

σ=2 for homonuclear diatomics

Equation 10.6 has the major merit that it allows us to use whatever value of the symmetry factor, σ, may be appropriate for the molecule in question. Non-linear molecules are also subject to rotational indistinguishability. Rotation of the H_2O molecule through half a revolution about its C_2 axis results in a molecule indistinguishable from the original. For NH_3, one third of a revolution about its C_3 axis has the same effect. Thus $\sigma(H_2O) = 2$ and $\sigma(NH_3) = 3$. Similar arguments assign to CH_4 and C_6H_6 the identical symmetry number of 12.

Table 10.1 summarises some themes of this and the preceding sections. Values of T/θ_r and q_{rot} are given at room temperature (300K).

Although, in general, the rotational temperature decreases as the mass of the molecule goes up, it is important to consider the role of the symmetry number before concluding that the rotational partition function necessarily follows suit. In particular, the rotational partition functions of hydrogen and methane are the same, despite a marked difference in rotational temperatures. Similarly, values for nitrogen and carbon monoxide are quite different, despite the closely similar masses and rotational temperatures of these molecules.

Table 10.1 Some rotational properties of molecules

	θ_r/K	σ	T/θ_r	q_{rot}
H_2	88	2	3.4	1.7
CH_4	15	12	20	1.7
HCl	9.4	1	32	32
HI	7.5	1	40	40
N_2	2.9	2	100	50
CO	2.8	1	110	110
CO_2	0.56	2	540	270
I_2	0.054	2	5600	2800

For diatomic species, I goes as μr^2 and θ_r goes as the inverse of this.

10.5 Origin of the symmetry factor

The symmetry factor has its origin in quantum mechanics and is related to the symmetry of the total wavefunction of the species in question. Under the interchange of identical particles (such as, say, the identical nuclei of a homonuclear diatomic molecule) the total wavefunction may be **symmetric** (SYM) (in which case its sign remains **unchanged**) or it may be

antisymmetric (AS) (in which case its sign **changes**). Atoms with integral nuclear spin are called **bosons** and their diatomic molecules have symmetric total wavefunctions. Those with half-integral nuclear spin are called **fermions** and their diatomic molecules have antisymmetric total wavefunctions. This information is summarised in Table 10.2.

Table 10.2 Symmetry features

Nucleus	Nuclear spin	Total wavefunction
Boson	integral	symmetric
Fermion	half-integral	antisymmetric

This overall symmetry requirement in diatomic molecules can best be understood if we consider all of the components that go to make up the total wavefunction, ψ_{tot}, and then consider the symmetry properties of each component in turn. We start by invoking the **Born–Oppenheimer** approximation which allows us to factorise the total wavefunction into separate components, one for each independent energy mode. We have met this idea already in Sections 9.2 and 9.4 where we factorised the various contributions to internal and external energy modes. In particular, we were able to write for the total energy (eqn 9.1)

$$\varepsilon_{tot} = \varepsilon_{trs} + \varepsilon_{rot} + \varepsilon_{vib} + \varepsilon_{el}.$$

Such a separation of the energy modes is coupled to the possibility of factorising the total wavefunction. We do this in eqn 10.7 and include all possible contributions in the total wavefunction in order to examine and reject any that have no influence on the overall symmetry under the interchange of identical nuclei

$$\psi_{tot} = \psi_{trs} \cdot \psi_{rot} \cdot \psi_{vib} \cdot \psi_{el} \cdot \psi_{ns} \quad (10.7)$$

The translational, rotational, vibrational, and electronic factors are already familiar. What is new is the **nuclear spin wave function,** ψ_{ns}. We will start with this least familiar aspect and then look at the other factors.

- ψ_{ns}: nuclear spin wavefunctions, whether in bosons or in fermions, can either be **symmetric** or **antisymmetric**. If the two nuclear spin states in a diatomic molecule are parallel, the nuclear spin wavefunction is symmetric to the interchange of identical nuclei; if the two nuclear spin states are paired, it is antisymmetric.
- ψ_{rot}: rotational wavefunctions can also be either symmetric or antisymmetric. Rotational states with even *J* have symmetric wavefunctions; those with odd *J* have antisymmetric wavefunctions. This is illustrated in Fig. 10.1, which shows the effect of an interchange of identical nuclei on the symmetry of some rotational wavefunctions. The reasons for this behaviour lie outside the scope of this text but they are explained in many elementary books on molecular spectroscopy, such as that by Barrow (1962).

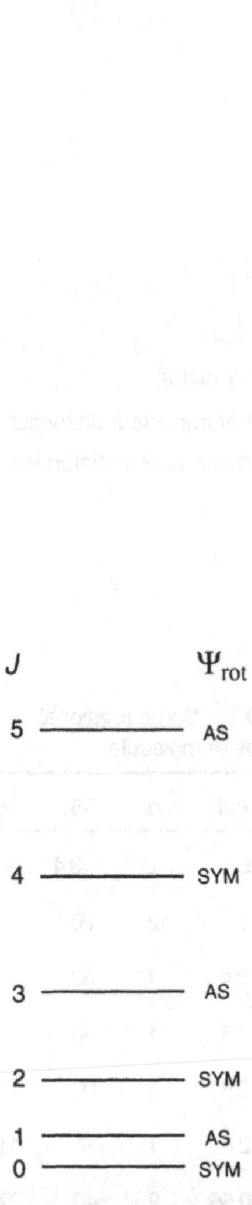

Fig, 10.1 Symmetry properties of the rotational wavefunction, ψ_{rot}, towards interchange of identical nuclei as a function of the rotational quantum number, *J*, in successive rotational energy levels of a homonuclear diatomic molecule.

- ψ_{el}: electronic wavefunctions, like those for nuclear spin and rotation, can also be either symmetric or antisymmetric. However, it is rare for a homonuclear diatomic molecule in its ground state to have other than a symmetric wavefunction. Most homonuclear diatomics have an electronic ground state, $^1\Sigma_g^+$, which is symmetric to the interchange of identical nuclei and can be ignored for the purposes of symmetry. Of the common homonuclear diatomic molecules, only molecular oxygen, O_2, has an electronic ground state that is antisymmetric, $^3\Sigma_g^-$. We shall return to this feature again in Section 11.4.

• $\psi_{\mathbf{trs}}$: translational wavefunctions depend only on the motion of the centre of mass of the molecule and hence have no effect on the overall symmetry. Translational wavefunctions are symmetric to the interchange of identical nuclei and can be ignored for the purposes of symmetry.

• $\psi_{\mathbf{vib}}$: vibrational wavefunctions depend only on the magnitude of the internuclear distance in the molecule and hence have no effect on the overall symmetry. Vibrational wavefunctions are symmetric to the interchange of identical nuclei and can also be ignored for the purposes of symmetry.

In trying to assess the effect of symmetry, we can eliminate all symmetric contributions to the total wavefunction (ψ_{trs}, ψ_{vib}, and, generally, ψ_{el}) from further consideration. They induce no symmetry concerns. What we are left with, then, is the product of just two wavefunctions, $\psi_{rot}\times\psi_{ns}$, as the determining elements of the overall symmetry.

The overall symmetry requirement in the total wavefunction can be satisfied only if the correct symmetry in the rotational wavefunction is coupled to the appropriate symmetry of the nuclear spin wavefunction. For example, in hydrogen, the nucleus has nuclear spin $I = \frac{1}{2}$, and is a **fermion** so the total wavefunction in the diatomic has to be antisymmetric. To achieve a total wave function that is antisymmetric, rotational states with **odd *J* values** must couple with antisymmetric nuclear spin wavefunctions (parallel spins), and rotational states with **even *J* values** must couple with the antisymmetric nuclear spin wavefunction (paired spins).

In deuterium, the situation is reversed. The deuteron has nuclear spin $I = 1$, and is a **boson**. So deuterium has a total wavefunction that is **symmetric**. Odd J values couple with the paired nuclear spin state and even J values with parallel spin states.

This is summarised in Table 10.3, and can be used to justify the inclusion of a symmetry factor of 2 for homonuclear diatomics. For if, in the summation of eqn 10.3, we have to run only over even J values or only over odd J values, then effectively we are counting only half the states in each case. Consequently, q_{rot} assumes only half the value it would reach were the summation to include all J values, as happens in a heteronuclear diatomic molecule.

Table 10.3 The coupling of nuclear and rotational states

	Proton	Deuteron
Nature	fermion	boson
Nuclear spin, I	$\frac{1}{2}$	1
Total wave-function	AS	SYM
Nuclear spin and rotation states	nuclear spins paired (AS) couple with even J (SYM)	nuclear spins paired (AS) couple with odd J (AS)
	nuclear spins parallel (SYM) couple with odd J (AS)	nuclear spins parallel (SYM) couple with even J (SYM)

10.6 The canonical partition function for rotation

As was seen in eqn 9.6, rotational energy is localised on a molecule so we can relate the canonical to the molecular partition function quite readily using eqn 10.6.

$$Q_{rot} = q_{rot}^N \tag{10.8}$$

Consequently, for the rotational canonical partition function, we have

$$Q_{rot} = \left[\frac{T}{\sigma\theta_r}\right]^N = \left[\frac{T}{\sigma hcB}\right]^N = \left[\frac{8\pi^2 IkT}{\sigma h^2}\right]^N \tag{10.9}$$

Toolkit functions

$$U = kT^2\left(\frac{\partial \ln Q}{\partial T}\right)_V$$

$$p = kT\left(\frac{\partial \ln Q}{\partial V}\right)_T$$

$$C_V = 2kT\left(\frac{\partial \ln Q}{\partial T}\right)_V + kT^2\left(\frac{\partial^2 \ln Q}{\partial T^2}\right)_V$$

$$S = k\ln Q + kT\left(\frac{\partial \ln Q}{\partial T}\right)_V$$

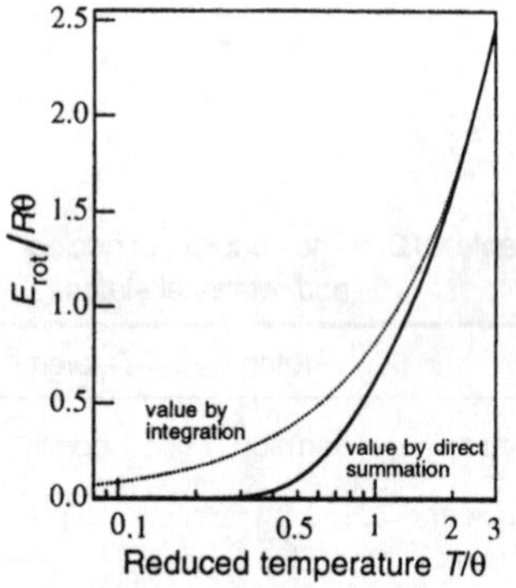

Fig. 10.2 The rotational energy using summation (eqn 10.3) or integration (eqn 10.4).

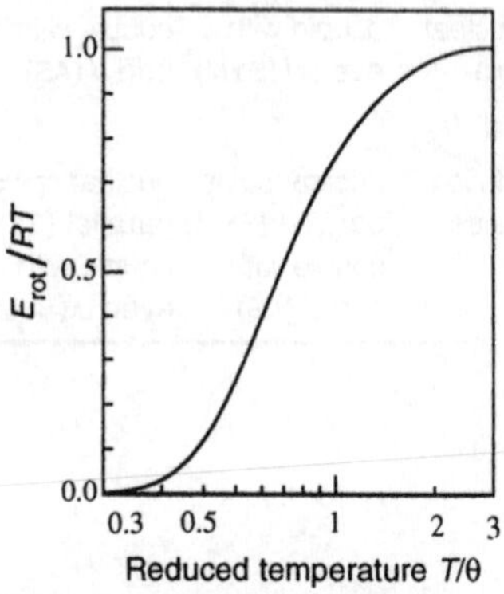

Fig. 10.3 The low temperature fall-off of rotational energy as E_{rot} falls below the classical value of RT.

10.7 Rotational energy, heat capacity, and entropy

The thermodynamic functions for rotation can be found using standard toolkit functions. We proceed in the same manner as we did in Chapter 8 for the translational energy, and start by taking the logarithm of eqn 10.9

$$\ln Q_{\text{rot}} = N\ln T + N\ln\left[\frac{8\pi^2 Ik}{\sigma h^2}\right] \tag{10.10}$$

which can be differentiated quite simply with respect to temperature, since the second term is a constant and has no temperature dependence

$$U_{\text{rot}} = kT^2\left(\frac{\partial \ln Q_{\text{rot}}}{\partial T}\right)_V = NkT^2\left(\frac{\partial}{\partial T}\ln T\right)$$

$$\therefore U_{\text{rot}} = NkT \quad \text{(for diatomic molecules)} \tag{10.11}$$

Equation 10.11, derived as it has been from a consideration of diatomic molecules, applies equally to all **linear molecules**, which have only two degrees of freedom in rotation. It can be recast for one mole of substance and, by taking the temperature derivative, can yield the molar rotational heat capacity, C_{rot}. Thus, when $N = L$, the **molar** rotational energy is $U_{\text{rot,m}}$

$$U_{\text{rot,m}} = RT \quad \text{and} \quad C_{\text{rot,m}} = R \text{ (linear molecules)} \tag{10.12}$$

None of these results is in any way new or unexpected; kinetic gas theory predicts $U_{\text{rot}} = RT$. But the exact way in which U_{rot} falls below the classical value of RT (Figs. 10.2 and 10.3) could only have been foretold from a quantum viewpoint.

As in the case of translational motion, it is the entropy equation that provides most of the new and unexpected information. By adapting the toolkit expression for entropy, as is done in eqn 10.13, we can write, for the rotational entropy

$$S_{\text{rot}} = kT\left(\frac{\partial \ln Q}{\partial T}\right)_V + k\ln Q = \frac{U_{\text{rot}}}{T} + k\ln Q$$

$$= \frac{NkT}{T} + k\ln\left[\frac{8\pi^2 IkT}{\sigma h^2}\right]^N$$

$$\therefore S_{\text{rot}} = Nk\left[1 + \ln\frac{IT}{\sigma} + \ln\left(\frac{8\pi^2 k}{h^2}\right)\right] \tag{10.13}$$

Again, as in the case of the translational entropy, we find an unpredicted dependence on mass. In this case, it is the **reduced mass**, μ, that is important. The mass dependence arises through the moment of inertia, I, which depends on the reduced mass ($I = \mu r^2$). There is also a constant (the final term). So, for the molar rotational entropy we have

$$S_{rot}/R = \ln\left[\left(I/\text{kg m}^2\right)(T/\text{K})(\sigma)^{-1}\right] + constant_4 \qquad (10.14)$$

where $constant_4$ takes the value 106.53.

Typically, q_{rot} at room temperature is of the order of hundreds for diatomics such as CO and Cl_2, which are not too light. Contrast this with the almost immeasurably larger value that the translational partition function reaches.

At room temperature $q_{rot} \approx 10^2$, but $q_{trs} \approx 10^{28}$

10.8 Extension to polyatomic molecules

In the most general case, that of a non-linear polyatomic molecule, there are three independent moments of inertia, one moment for each mutually perpendicular axis of rotation. The rotational partition function has to take account of each of these three moments of inertia, and one way of doing so is by recognising the existence of three independent characteristic rotational temperatures $\theta_{r,x}$, $\theta_{r,y}$, and $\theta_{r,z}$, corresponding to the three principal moments of inertia I_x, I_y, and I_z.

$$\theta_r = \frac{h^2}{8\pi^2 I k} = \frac{hcB}{k}$$

The resulting expression (eqn 10.15) is very similar in overall form to that in eqn 10.9.

$$q_{rot} = \frac{\sqrt{\pi}}{\sigma}\left[\left(\frac{T}{\theta_{r,x}}\right)\left(\frac{T}{\theta_{r,y}}\right)\left(\frac{T}{\theta_{r,z}}\right)\right]^{\frac{1}{2}} \qquad (10.15)$$

Using eqn 10.15, the partition function, and hence all the important thermodynamic expressions involving molecular rotation in a non-linear polyatomic molecule, can be determined.

10.9 Conclusions

Rotational energy levels, although more widely spaced than translational energy levels, are still close enough at most temperatures to allow us to use the **continuum approximation** and to replace the summation in q_{rot} by an integration. Providing proper regard is then paid to **rotational indistinguishability**, as shown through considerations of symmetry, calculated rotational thermodynamic functions are entirely as expected.

The rotational symmetry factor can be linked to the symmetry properties of the total wavefunction and hence, via the nuclear spin, to the **boson** or **fermion** nature of atomic nuclei. The implications of this point are developed further in the next chapter.

Reference

Barrow, G. M. (1962). *Introduction to Molecular Spectroscopy*. McGraw-Hill, New York.

11 *ortho* and *para* spin states: a case study

11.1 Introduction

It seems appropriate, before leaving the topic of rotational energies, to consider the questions raised in the case of *ortho*- and *para*-hydrogen or deuterium. We laid the basic groundwork for this in Section 10.3, where we discussed the coupling of odd J and even J rotational states with parallel and paired nuclear spin states. All that remains now is to determine how many nuclear spin states contribute to the total wavefunction and hence to find the statistical weight of the odd J and even J rotational states in a sample of hydrogen or deuterium.

Table 11.1 *o*- and *p*- states

Nuclear spin	I		
Spin states per nucleus	ρ	=	$2I+1$
Spin states per molecule	ρ^2	=	$(2I+1)^2$
para states (SYM)	$\frac{1}{2}\rho(\rho-1)$	=	$I(2I+1)$
ortho states (AS)	$\frac{1}{2}\rho(\rho+1)$	=	$(I+1)(2I+1)$

11.2 Nuclear spin wavefunctions

In general, for a homonuclear diatomic molecule with nuclear spin I, each nucleus can have ρ $(= 2I + 1)$ spin states, and a total of ρ^2 nuclear spin wavefunctions to include in the total wavefunction for the diatomic molecule as a whole. Of these ρ^2 states, $\frac{1}{2}\rho(\rho + 1)$ comprise symmetric (SYM) nuclear spin states and $\frac{1}{2}\rho(\rho - 1)$ comprise antisymmetric (AS) nuclear spin states. By convention, we call the states with higher multiplicity, $\frac{1}{2}\rho(\rho + 1)$, *ortho* states and the others *para* states. Thus, **ortho states** are associated with **symmetric** nuclear spin wavefunctions and **para states** are associated with **antisymmetric** nuclear spin wavefunctions. This remains true whether the two nuclei are **bosons** or **fermions**.

These conclusions are summarised in Table 11.1. In Table 11.2, the information in Table 11.1 is developed for two specific cases, hydrogen and deuterium.

Table 11.2 Hydrogen and deuterium

	Hydrogen	Deuterium
Nuclear spin, I	$\frac{1}{2}$	1
Spin states per nucleus,	2	3
Spin states per molecule, ρ^2	4	9
ortho states (SYM) couple with	odd J	even J
para states (AS) couple with	even J	odd J
o-/*p*- ratio (high T)	3 :1	2:1

The **proton** is a **fermion** with $I = \frac{1}{2}$ and a total wavefunction that is **antisymmetric** for the exchange of the identical fermion nuclei. There are two spin states per nucleus (up, down) and four nuclear spin wavefunctions to include in the total wavefunction. Of these, three are *ortho* states with symmetric nuclear spin wavefunctions which couple with rotational states having odd J, and one *para* state with an antisymmetric nuclear spin wavefunction which couples with rotational states having even J.

The **deuteron** is a **boson** with $I = 1$ and a total wavefunction that is **symmetric** for the exchange of the identical boson nuclei. There are three spin states per nucleus and nine nuclear spin wavefunctions to include in the total wavefunction. Of these, six are *ortho* states with **symmetric** nuclear spin wavefunctions; these couple only with rotational states which have

even *J* values. The remaining three are *para* states with **antisymmetric** nuclear spin wavefunctions; these couple only with rotational states which have **odd *J*** values.

Thus *o*-hydrogen comprises only odd rotational states, whereas *o*-deuterium comprises only even rotational states. Moreover, because the statistical weight of *ortho* states is greater than that of *para* (by a factor of 3 for hydrogen and by a factor of 2 for deuterium), the relative intensities of adjacent rotational lines in the **rotational Raman** spectrum are in the ratio 3:1 for hydrogen with **odd** lines predominating, and in the ratio 2:1 for deuterium, where **even** lines predominate. Figure 11.1 illustrates the origin of these alternating intensities and may be compared with Fig. 10.1, from which it takes its origin.

Rotational Raman spectrum: The existence of rotational states in homonuclear diatomics cannot be observed directly since symmetric molecules with identical nuclei have no dipole moment and so cannot absorb in the infrared, to give a direct rotational spectrum. Rotational Raman spectra do not require a dipole moment.

Alternating intensities of spectral lines.

The formal way of writing the partition function for rotation in hydrogen is

$$q_{\mathrm{rot}}^{\mathrm{H}_2} = \tfrac{1}{4}\left\{\sum_{\textit{even } J}(2J+1)\mathrm{e}^{-J(J+1)\theta_r/T} + 3\sum_{\textit{odd } J}(2J+1)\mathrm{e}^{-J(J+1)\theta_r/T}\right\} \quad (11.1)$$

with the equivalent expression for deuterium being

$$q_{\mathrm{rot}}^{\mathrm{D}_2} = \tfrac{1}{3}\left\{2\sum_{\textit{even } J}(2J+1)\mathrm{e}^{-J(J+1)\theta_r/T} + \sum_{\textit{odd } J}(2J+1)\mathrm{e}^{-J(J+1)\theta_r/T}\right\} \quad (11.2)$$

Both these equations give the same limiting expression if the summations extend over a large number of terms, as is true at high temperatures when $T/\theta_r \gg 1$. The high temperature limiting expression is

$$q_{\mathrm{rot}} = \tfrac{1}{2}\sum_{\textit{all } J}(2J+1)\mathrm{e}^{-J(J+1)\theta_r/T} \quad (11.3)$$

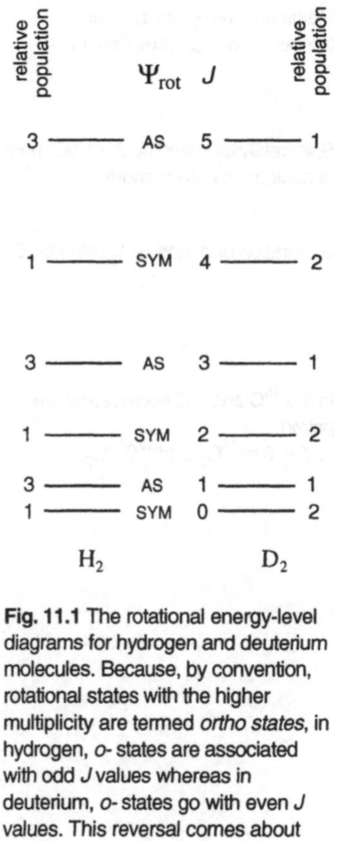

Fig. 11.1 The rotational energy-level diagrams for hydrogen and deuterium molecules. Because, by convention, rotational states with the higher multiplicity are termed *ortho states*, in hydrogen, *o*- states are associated with odd *J* values whereas in deuterium, *o*- states go with even *J* values. This reversal comes about because the proton is a *fermion*, whereas the deuteron is a *boson*.

As always, *high temperature* is a relative term which relies on a comparison of the actual temperature with the rotational characteristic temperature θ_r. For all but the isotopes of hydrogen (H_2, D_2, and HD), the rotational temperature is less than one-tenth of room temperature, so the approximation introduced by eqn 11.3 can be ignored with confidence.

Equation 11.3 gives further confidence to the introduction of the **symmetry factor**, $\sigma = 2$, for homonuclear diatomics and, by inference, to the use of other values of σ for molecules with higher rotational symmetry (see, for example, Section 10.3). In some cases, it is the symmetry factor that can have an overriding influence on the calculated value of certain thermodynamic functions.

It should be noted that σ enters expressions for entropy, *S*, and for the Helmholtz and Gibbs functions, *A* and *G*, but is absent from expressions for the internal energy, *U*, and also from the heat capacity, C_V.

11.3 *ortho*- and *para*- hydrogen

At high temperatures, hydrogen exists as an equilibrium mixture of spin states with *o*-hydrogen enjoying a 3:1 predominance. As the temperature is lowered, however, the natural tendency of molecules to drop into the lowest rotational energy level ($J = 0$) asserts itself. But, because $J = 0$ belongs to the set of **even *J*** values, this tendency implies the need to convert the nuclear spin state of *o*-hydrogen (parallel nuclear spins) to that of *p*-hydrogen (paired nuclear spins).

Spin conversion of this kind can occur, but is always rather a **slow process**, particularly as the temperature is lowered. So the 3:1 *o*- to *p*- ratio can persist (can be *frozen-in*) in samples of hydrogen down to room temperature and below.

Metastable n-H_2 has the high temperature *o*-:*p*- composition.

The metastable 3:1 mixture is called **normal hydrogen** (n-H_2) and a mixture with this composition can indeed exist right down to the absolute zero, with $\frac{3}{4}$ of the molecules in the *ortho* state and only able to drop their rotational energy to the $J = 1$ level. Thus, the vast majority of the molecules in liquid n-H_2 are not in their lowest rotational energy state ($J = 0$) and can be said to have what amounts to a finite rotational zero-point energy.

Spin catalysts speed up the attainment of nuclear spin equilibrium.

Separation of *o*- and p-H_2 using GSC.

The slow interconversion of nuclear spin states can be accelerated by using a **spin catalyst** (usually a paramagnetic salt or the process of sorption on a metal or activated charcoal surface). It is possible to prepare a sample of pure *p*-hydrogen by this means. Removal of the spin catalyst leaves pure, metastable, *p*-hydrogen which resists rapid conversion to a mixture of *o*- and *p*-hydrogen at temperatures well up to room temperature and above. Pure *o*-hydrogen can be prepared using gas-solid chromatography (GSC) on n-H_2. It, too, resists spin interconversion even at quite high temperatures unless a spin catalyst is introduced.

11.4 A special case: nuclei with zero-spin

In the ^{16}O and ^{12}C nuclei, spins are paired
$\therefore I = 0$ in $^{16}O_2$ and $^{12}C^{16}O_2$.

Some nuclei have zero nuclear spin, $I = 0$. For example, the ^{16}O nucleus has zero spin. The eight protons and eight neutrons each occupy their own nuclear energy manifolds which, with eight fundamental particles present, each form closed-shell configurations with all like nuclear spins paired. The same is true in ^{12}C, another closed-shell nuclear configuration which results in a nuclear spin $I = 0$.

The condition $I = 0$ implies that the number of spin states for the diatomic molecule, or for the linear triatomic like CO_2, is only 1 ($\rho = 2I + 1 = 1$ as also is $\rho^2 = 1$). So, since $I = 0$ is an integral spin, $^{16}O_2$ and CO_2 have oxygen (and carbon) nuclei that are bosons and thus require a total wavefunction that is **symmetric**.

In the rotational Raman spectrum of CO_2, all the odd *J* lines are missing.

It is simplest to start by discussing the case of carbon dioxide. As there is only one nuclear spin state to consider, only one rotational spin state can couple with it. The symmetric nuclear spin wavefunction can only couple with **even *J*** values if the outcome is to be symmetric. All the transitions in the rotational Raman spectrum of CO_2 which involve odd *J* states are missing and only the even *J* lines appear in the spectrum.

In $^{16}O_2$, all the even *J* lines are missing.

In the $^{16}O_2$ molecule, a similar line of reasoning leads to the conclusion that half the expected lines in the rotational Raman spectrum will be missing. But this time, it is the lines of even *J* that are absent. The reason for this apparent reversal lies in the fact that the ground electronic state in oxygen, ${}^3\Sigma_g^-$, is antisymmetric, so the overall requirement for a symmetric total wavefunction is met by coupling a symmetric nuclear spin state to an antisymmetric ground electronic state together with an antisymmetric set of rotational energy levels which have odd J values only.

The energy-level diagrams for $C(^{16}O)_2$ and for $^{16}O_2$ are shown together in Fig. 11.2.

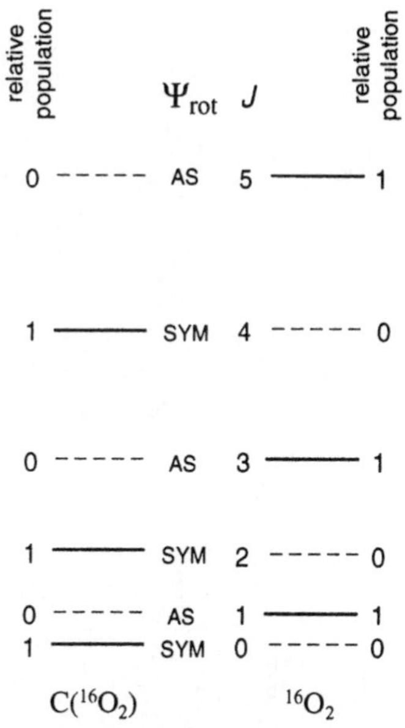

Fig, 11.2 The rotational energy-level diagrams for carbon dioxide and oxygen molecules, both containing only the ^{16}O isotope of oxygen. Alternate lines in each spectrum are missing because ^{16}O has nuclear spin $I = 0$ and, consequently, only one spin state. The reversal of lines between the two species arises because the electronic ground state has a symmetric wavefunction in CO_2, but an antisymmetric one in O_2.

11.5 Conclusions

Nuclear spin energy states, which depend on the various orientations of the nuclear spin, are very closely spaced in energy. Such small energy separations make the nuclear spin characteristic temperature also very small, so it is rare for nuclear spin states to act in any way other than classically at any reasonable (low) experimental temperature. Only in hydrogen and its isotopes at accessible low temperatures does the equilibrium orientation of nuclear spin vary significantly from the random expectation, so it is only for the isotopes of hydrogen that nuclear spin needs to be considered in any study of the heat capacity, and then only at very low temperatures.

The entropy associated with differing orientations of any given nucleus with a non-zero spin remains unchanged by bonding to other species and is rarely altered by a change in temperature. Consequently, it is quite safe to ignore the effects of nuclear spin in all practical considerations of chemical change.

The importance of nuclear spin is that it determines the relative populations (statistical weights) of successive rotational states in homonuclear and other symmetrical molecules. This is the realm of what is called **nuclear statistics**, which must be considered on every occasion when the rotation of a molecule results in the interchange of identical nuclei. The coupling of *ortho* and para nuclear spin states to odd and even rotational J values enables us to provide an account of alternating intensities (and sometimes missing intensities) of lines in the rotational Raman spectra of isolated molecules even at room temperature and above. Nuclear spin also provides an insight into the existence of *ortho* and *para* states and into the low-temperature thermal behaviour in hydrogen and its isotopes.

12 The ideal diatomic gas: the vibrational partition function

12.1 Introduction

Vibrational modes have energy level spacings that are larger by at least an order of magnitude than those in rotational modes, which, in turn, are some 25 – 30 orders of magnitude larger than those of the translational modes. Consequently, the treatment of vibrational modes cannot, on the one hand, be simplified by using the continuum approximation, but neither, on the other hand, are they going to enjoy appreciable excitation at room temperature anyway.

It is often safe to assume, therefore, that the value of the vibrational partition function, q_{vib}, at 300 K is imperceptibly different from unity. This approximation can be applied quite rigorously to light molecules (e.g. H_2, D_2, CH_4, NH_3, N_2, CO, O_2), but is already some 10% in error for more complicated molecules such as CO_2.

12.2 The diatomic SHO model

We start by modelling a diatomic molecule on a simple **ball and spring** basis with two atoms, mass m_1 and m_2, joined by a spring which has a force constant k.

The classical vibrational frequency, ν (simple harmonic oscillator, SHO, model), is given by

$$\nu = \frac{1}{2\pi}\sqrt{\frac{k}{\mu}} \tag{12.1}$$

where μ is the reduced mass, which was introduced for rotation in Section 10.2.

There is a quantum restriction on the available energies

$$\varepsilon_{vib} = \left(v + \tfrac{1}{2}\right)h\nu \tag{12.2}$$

$\uparrow$ zero-point energy

where the quantum number v is an integer and can take values any positive value from zero upwards.

Vibrational energy levels in diatomic molecules, unlike rotational levels, are always **non-degenerate**. Degeneracy of vibrational modes does, however, have to be taken into consideration in polyatomic species, as indicated below.

force constant k

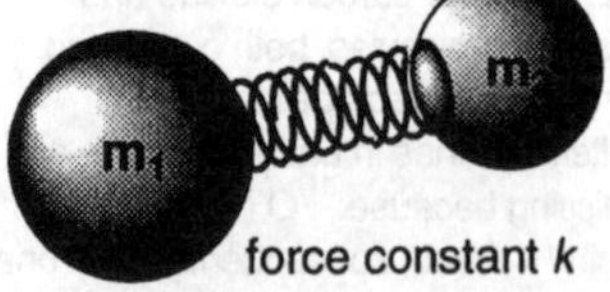

Ball and spring model

The vibrational potential well is not strictly harmonic, especially for highly excited vibrational states. Corrections for **anharmonicity** can be made,

$\varepsilon_{vib} = (v + \tfrac{1}{2})h\nu - \chi_e(v + \tfrac{1}{2})^2 h\nu$

χ_e is the vibrational **anharmonicity constant.** However, thermal excitation of any but the lowest vibrational states requires such high temperatures that this correction is rarely needed.

Polyatomic species

Polyatomic molecules can undergo many independent vibrational motions. For these, **normal mode analysis** allows us to assume that each **normal mode** behaves as an independent harmonic oscillator to which the SHO model may be applied.

In general, a polyatomic molecule containing N atoms will have $(3N - 5)$ normal modes of vibration if it is **linear** and $(3N - 6)$ if it is **non-linear**.

Normal mode
An independent, synchronous motion of atoms or groups of atoms that may be excited without leading to the excitation of any other normal mode.

Vibrational degrees of freedom
An assembly of N uncoupled atoms free to move about independently will have a total $3N$ (translational) degrees of freedom. If the atoms are then coupled (by chemical bonds), to form an N-atom molecule, the total number of degrees of freedom remains unchanged at $3N$, but some of these modes are now assigned to rotation and vibration. Three modes go to translation of the molecule as a whole, and a further three (two if the molecule is linear) to rotation. This leaves $(3N - 6)$ modes which can be assigned to normal modes of vibration (if the molecule is non-linear) and $(3N - 5)$ normal modes of vibration if the molecule is linear.

12.3 The vibrational partition function, q_{vib}

We start by making a simplifying assumption and set $\varepsilon_0 = 0$, that is we set the ground vibrational state as the reference zero for vibrational energy. We then measure all other energies relative to this arbitrary but sensible zero reference, ignoring the zero-point energy. This means that in calculating values of some vibrational thermodynamic functions (notably the vibrational contribution to the internal energy, U) we will have to add the sum of the individual zero-point energies of all normal modes present. The need to remember this is a small price to pay for the simplification it affords.

This assumption ($\varepsilon_0 = 0$) allows us to write

$$\varepsilon_1 = h\nu, \quad \varepsilon_2 = 2h\nu, \quad \varepsilon_3 = 3h\nu, \quad \varepsilon_4 = 4h\nu, \ \ldots, \text{ etc.}$$

Under this assumption, q_{vib} may be written as

$$q_{vib} = \sum e^{-\beta\varepsilon_{vib}} = 1 + e^{-\beta h\nu} + e^{-2\beta h\nu} + e^{-3\beta h\nu} + \ldots \qquad (12.3)$$

a simple geometric series which yields q_{vib} in closed form as

$$q_{vib} = \frac{1}{1 - e^{-\beta h\nu}} = \frac{1}{1 - e^{-\theta_{vib}/T}} \qquad (12.4)$$

where θ_{vib}, the **characteristic vibrational temperature**, is defined alongside. Unlike the situation for rotation, θ_{vib} can be identified with an actual separation between quantised energy levels. To a very good degree of approximation, since the anharmonicity correction can be neglected for low quantum numbers, the vibrational temperature is characteristic of the gap between the lowest and first excited vibrational states, and with *exactly twice* the zero-point energy, $\frac{1}{2}h\nu$.

$$\theta_{vib} = \frac{h\nu}{k}$$

Equation 12.4 applies to diatomic molecules and can also be applied to polyatomic species with each normal mode of vibration assumed to be independent and taken separately.

Vibrational energy spacings are much larger than those for rotation, so typical characteristic vibrational temperatures in diatomic molecules are of the order of hundreds to thousands of kelvins rather than the tens to hundreds of kelvins characteristic of rotation. Tables 12.1 and 12.2 show values of θ_{vib} and of q_{vib} at 300 K for some typical diatomic and polyatomic species.

In light diatomic molecules, which have high force constants as well as low reduced masses, vibrational frequencies (eqn 12.1) and characteristic vibrational temperatures are high. Consequently, there is only one vibrational state, the ground state, that is accessible to these species at room temperature, and the vibrational partition function has a value that is never

Table 12.1 Characteristic vibrational temperatures for diatomic molecules

Species	θ_{vib}/K	q_{vib} at 300 K
H_2	5987	1.000
HD	5226	1.000
D_2	4307	1.000
N_2	3352	1.000
CO	3084	1.000
Cl_2	798	1.075
I_2	307	1.556

far from unity. In contrast, heavy diatomics such as iodine have rather a loose vibration with a lower characteristic temperature. For such species, even at room temperature, there is appreciable vibrational excitation and consequent population of the first and, to a slight extent also, of higher excited vibrational energy states.

The situation in polyatomic species is similar, complicated only by the existence of $3N - 5$ or $3N - 6$ normal modes of vibration. Commonly, some of these normal modes are degenerate, as can be seen in Table 12.2. The final column in Table 12.2 gives the overall value of the vibrational partition function as the product of mutually independent contributions from individual normal modes according to

$$q_{vib}^{tot} = \prod\left(q_{vib}^{(n)}\right) = q_{vib}^{(1)} \times q_{vib}^{(2)} \times q_{vib}^{(3)} \times \ldots \quad (12.5)$$

with (1), (2), and (3), ..., denoting individual normal modes 1, 2, 3, ... etc. This procedure is justified to the extent that the normal mode energies are **independent** and **factorisable**. This will generally be so if the effects of **anharmonicity** can be neglected, which will be true at all but the very highest temperatures.

As with diatomics, only the heavier species show values of q_{vib} appreciably different from unity, thus indicating some population at 300 K of states lying above the ground state for such molecules but no appreciable equivalent population for lighter molecules.

Typically, then, θ_{vib} is of the order of ~3000 K in many molecules. Consequently, at 300 K we have (eqn 12.4)

$$q_{vib} = \frac{1}{1 - e^{-10}} \approx 1$$

in marked contrast with q_{rot} (≈ 10) and q_{trs} ($\approx 10^{30}$). Most molecules have only one accessible state, the ground state, in vibration.

Table 12.2 Characteristic vibrational temperatures in polyatomic molecules. Degeneracies are shown in brackets after the corresponding temperatures.

Species	θ_{vib}/K	$\prod(q_{vib})$ at 300 K
CO_2	3360	1.091
	1890	
	954 (2)	
NH_3	4880 (2)	1.001
	4780	
	2330 (2)	
	1360	
H_2O	5360	1.000
	5160	
	2290	
$CHCl_3$	4330	2.650
	1745 (2)	
	1090 (2)	
	938	
	523	
	374 (2)	

12.4 High temperature limiting behaviour of q_{vib}

At high temperatures, eqn 12.4 gives, to a good approximation, a linear dependence of q_{vib} with temperature. If we expand $1 - e^{-\theta_{vib}/T}$, we get

$$q_{vib} \approx \frac{1}{1 - 1 + \left(\theta_{vib}/T\right) + \ldots} = \frac{T}{\theta_{vib}} \quad \text{(high temperature limit)} \quad (12.6)$$

In Fig. 12.1, the temperature variation of the vibrational partition function is shown. As the temperature increases, the linear dependence of q_{vib} upon T becomes increasingly clear.

12.5 The canonical partition function, Q_{vib}

As is usual,

$$Q_{vib} = q_{vib}^{N} = \left(\frac{1}{1 - e^{-\theta_{vib}/T}}\right)^{N} \quad (12.7)$$

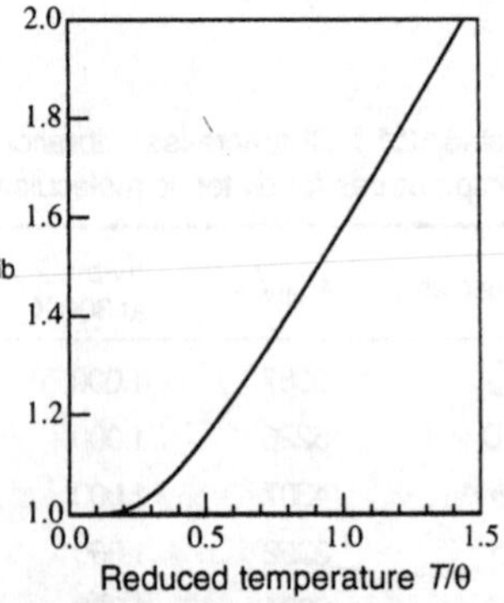

Fig. 12.1 The temperature dependence of the vibrational partition function. q_{vib} becomes a linear function of T once $T \gg \theta_{vib}$.

so, using the appropriate toolkit function, we can find the first differential of $\ln Q$ with respect to temperature, to give

$$U_{\text{vib}} = kT^2\left(\frac{\partial \ln Q_{\text{vib}}}{\partial T}\right)_V = \frac{Nk\theta_{\text{vib}}}{\left(e^{\theta_{\text{vib}}/T} - 1\right)} \quad (12.8)$$

Toolkit function

$$U = kT^2\left(\frac{\partial \ln Q}{\partial T}\right)_V$$

12.6 Vibrational energy, U_{vib}

The molar vibrational energy, from eqn 12.8, is thus

$$U_{\text{vib,m}} = \frac{R\theta_{\text{vib}}}{\left(e^{\theta_{\text{vib}}/T} - 1\right)} \quad (12.9)$$

which is not nearly as simple as

$$U_{\text{trs,m}} = \tfrac{3}{2}RT \quad (8.10)$$

for translational kinetic energy, nor as simple as

$$U_{\text{rot,m}} = RT \quad \text{(linear molecules)} \quad (10.11)$$

but which does, as we saw in Section 12.4, reduce to the simple and expected form at **equipartition** (at very high temperatures) giving

$$U_{\text{vib,m}} = RT \quad \text{(for each normal mode of vibration)} \quad (12.10)$$

Equipartition of energy occurs at temperatures high enough ($\Delta\varepsilon \gg kT$) for very many states to be occupied and for the system to behave classically.

More typically, at room temperature

$$U_{\text{vib,m}} = \frac{3000R}{\left(e^{10} - 1\right)} \approx \frac{1}{7}R$$

instead of the $300R$ predicted for equipartition.

12.7 The zero-point energy

So far, we have chosen the zero-point energy ($\tfrac{1}{2}h\nu$) as the zero reference of our energy scale. If we choose not to measure energies from the zero-point level but, instead, set $\varepsilon_0 = \tfrac{1}{2}h\nu$, we choose as a zero reference the rotational energy at the absolute zero, $T = 0$ K. This means that, to each term in the energy ladder, we need to add an amount $\tfrac{1}{2}h\nu$. In turn, this means that for each particle we add this same amount ($\tfrac{1}{2}h\nu$). Thus, for N particles, we must add an amount $U(0)_{\text{vib,m}} = \tfrac{1}{2}Nh\nu$, so

$$\begin{aligned} U_{\text{vib,m}} &= \frac{R\theta_{\text{vib}}}{\left(e^{\theta_{\text{vib}}/T} - 1\right)} + U(0)_{\text{vib,m}} \\ &= \frac{R\theta_{\text{vib}}}{\left(e^{\theta_{\text{vib}}/T} - 1\right)} + \frac{1}{2}Lh\nu \end{aligned} \quad (12.11)$$

Although $U(0)_{\text{vib,m}}$ is specifically chosen to take account of the zero-point energy, we could equally properly refer all our vibrational energies to any other reference zero of energy.

12.8 Vibrational heat capacity, C_{vib}

The vibrational heat capacity can be found by putting the partition function into the appropriate toolkit function, but it can also be found, more simply now that we have an explicit expression (eqn 12.11) for the vibrational energy relative to any chosen zero reference, by differentiating $U_{\text{vib,m}}$ directly with respect to temperature. The result is always independent of the chosen zero reference since its influence is not a function of temperature.

$C_{\text{vib,m}}$ is the **molar** heat capacity and applies to one mole of particles.

$$C_{\text{vib,m}} = \left(\frac{\partial U_{\text{vib,m}}}{\partial T}\right)_V = R\left(\frac{\theta_{\text{vib}}}{T}\right)^2 \frac{e^{\theta_{\text{vib}}/T}}{(e^{\theta_{\text{vib}}/T}-1)^2} \quad (12.12)$$

THE EINSTEIN EQUATION

The Einstein equation can be written in a more compact form as

$$C_{\text{vib,m}} = R\,\mathscr{F}_{\text{E}}\left(\frac{\theta_{\text{vib}}}{T}\right) \quad (12.13)$$

The Einstein function, $\mathscr{F}_{\text{E}}$, (read as F of E).

where $\mathscr{F}_{\text{E}}$, with argument θ_{vib}/T, is the **Einstein function**, which is tabulated in many standard handbooks as

$$\mathscr{F}_{\text{E}} = \frac{u^2 e^u}{(e^u-1)^2} \qquad \text{with} \quad u = \frac{\theta_{\text{vib}}}{T} \quad (12.14)$$

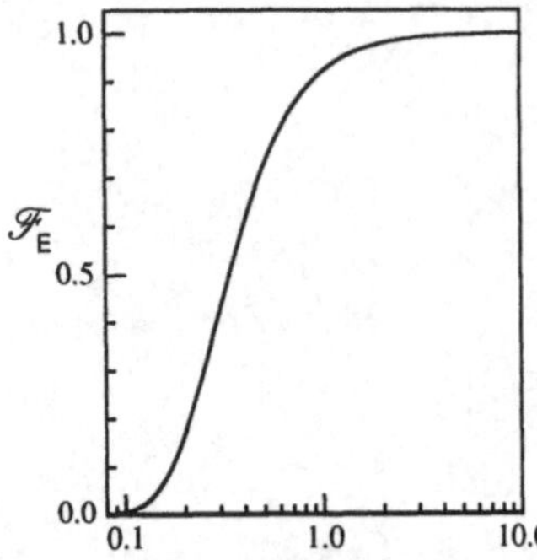

Fig. 12.2 The Einstein heat capacity. The logarithmic reduced temperature scale helps to highlight both high and low temperature behaviour.

The Einstein function has applications beyond normal modes of vibration in gas molecules. It also has an important place in the understanding of the influence of lattice vibrations (**phonons**) on the thermal behaviour of solids and is at the heart of one of the earliest (1909) models for the heat capacity of solids and allows the simplest explanation of the "fall-off" curve (Fig. 12.2) of heat capacities in solids, which remained both unexplained and inexplicable until the advent of quantum theory.

The Einstein function tends towards zero at low temperatures (when $\theta \gg T$) and reaches the limiting value of unity at high temperatures (when $\theta \ll T$). In this upper limit, behaviour is entirely classical and $C_{\text{vib,m}}$ takes the expected equipartition value for a harmonic oscillator, which is R.

12.9 The vibrational entropy, S_{vib}

Toolkit function

$$S = -\left(\frac{\partial A}{\partial T}\right)_V = -k\left(\frac{\partial(-T\ln Q)}{\partial T}\right)_V$$
$$= k\ln Q + kT\left(\frac{\partial \ln Q}{\partial T}\right)_V$$

From the molecular partition function for vibration, q_{vib}, we can deduce the canonical partition function, Q_{vib}, and, using the appropriate toolkit function, it is a relatively straightforward matter to derive an expression for S_{vib}. However, because we have already derived an expression for U_{vib}, it is an even simpler task to take a short-cut and use the classical expression for the entropy together with the Massieu function $A/T = -kT \ln Q$

$$S_{\text{vib}} = \frac{U_{\text{vib}} - U_{\text{vib}}(0)}{T} - \frac{A_{\text{vib}} - A_{\text{vib}}(0)}{T} = \frac{U_{\text{vib}} - U_{\text{vib}}(0)}{T} + k \ln Q_{\text{vib}}$$

For U_{vib} we use eqn 12.11 and subtract the zero-point energy, $U_{\text{vib}}(0) = \frac{1}{2}Lh\nu$ since it has no influence on the magnitude of the entropy. And for Q_{vib}, we use $Q_{\text{vib}} = q_{\text{vib}}^N$, with $N = L$ for one mole of oscillators. Consequently, we can write (again for one mole)

$$\ln Q_{\text{vib, m}} = Lk \ln q_{\text{vib}} = R \ln q_{\text{vib}} \quad (12.15)$$

The outcome of this is shown in eqn 12.16 and is plotted on a logarithmic reduced temperature scale in Fig. 12.3.

$$\frac{S_{\text{vib,m}}}{R} = \frac{\theta_{\text{vib}} / T}{\left(e^{\theta_{\text{vib}}/T} - 1\right)} - \ln\left(1 - e^{-\theta_{\text{vib}}/T}\right) \quad (12.16)$$

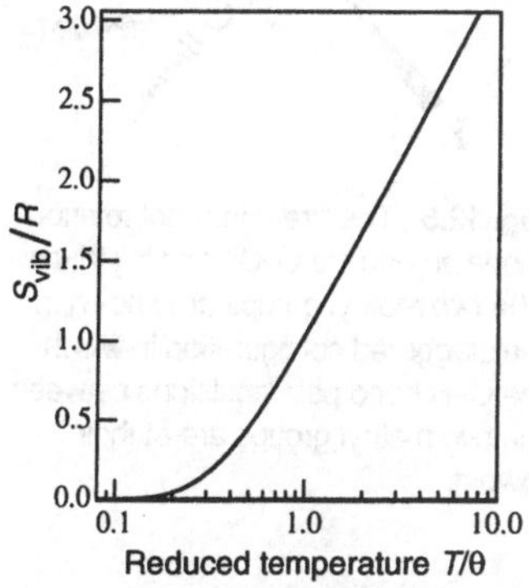

Fig. 12.3 The variation of the vibrational entropy with the reduced temperature T/θ_{vib}. Note the approach to zero entropy as $T \to 0$, and the unlimited increase in entropy as $T \to \infty$.

From Fig. 12.3, it is clear that the contribution to the overall entropy from vibrational modes is rarely large, unless the temperature begins to approach the characteristic vibrational temperature, θ_{vib}. As θ_{vib} is generally rather large, the vibrational entropy of most gases at room temperature can usually be ignored with good justification. The vibrational entropy tends to zero as the temperature approaches zero and tends to increase without limit as the temperature approaches infinity.

It is worth noting, however, that even moderately complicated molecules will have quite large numbers $(3N - 6)$ of vibrational modes and that the total contribution of these to the entropy can sometimes be appreciable. Thus a five-atom molecule (methanoic acid, for example) has nine normal modes thus increasing the vibrational entropy by almost 10 over a single mode, while a nine-atom molecule (ethanol, for example) has 27 normal modes which will tend to increase the vibrational entropy by a factor close to thirty

It is never safe to ignore the contributions from vibration until both the number of normal modes and the magnitude of the vibrational temperature for each mode has been explored.

12.10 From torsional oscillation to internal rotation

So far, we have considered only those types of internal motion that can clearly be identified either as "free rotation" or as "linear vibration". It is worth noting, however, that other intermediate forms of rotational/vibrational motion can exist within a molecule, and that the change of one into the other can affect profoundly both the value of the partition function and its temperature dependence, and also the values of all the thermodynamic parameters that depend critically on the partition function.

The problem posed is too specialised to be dealt with in an introductory text of this nature, so all we will attempt to do here is to outline the nature of the problem without attempting to provide any detailed or quantitative solutions.

This is best approached by considering the molecule ethene. This six-atom molecule has $(3N - 6) = 12$ normal modes. Eleven of these modes can be

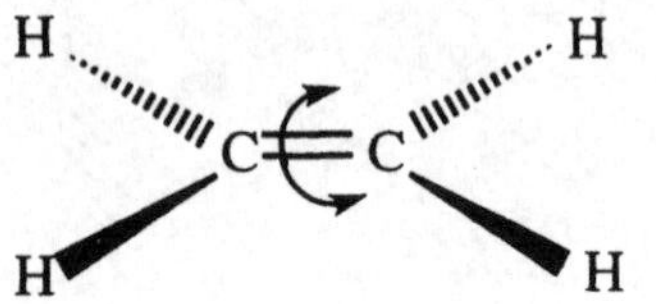

Fig. 12.4 The twisting vibrational mode around the C=C double bond in ethene.

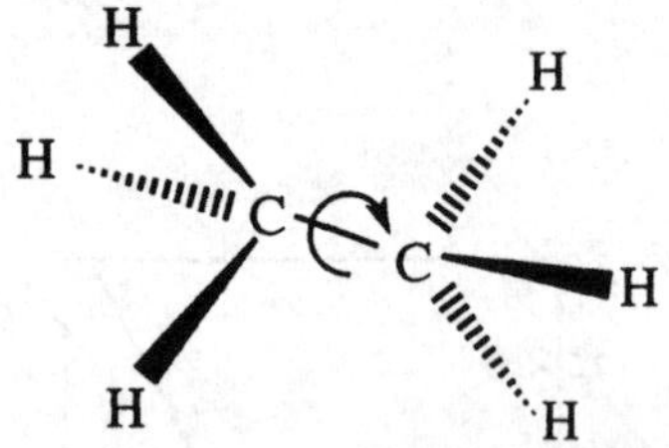

Fig. 12.5 The "free" internal rotation mode around the C–C bond in ethane. The two methyl groups are shown in the staggered configuration in which the C–H bond pair repulsions between the two methyl groups are at their lowest.

assigned spectroscopically and comprise conventional vibrational modes. The twelfth, which is spectroscopically inactive, is the **twisting mode** around the double bond, as shown in Fig. 12.4. There is no free rotation about the C=C double bond, but the two CH_2 groups can undergo a **torsional oscillation** to and fro about the C–C axis. This torsional oscillation is indistinguishable from a normal vibration, apart from being spectroscopically inactive.

Contrast this with the situation in ethane (Fig. 12.5). Here, the two methyl groups can rotate reasonably freely through 360° as the C–C sigma bond does not impose the same rotational rigidity as the C=C π-bond. However, rotational motion is not completely "free". In the eclipsed position, C–H bond pair repulsions between the two methyl groups are at their highest, whereas in the staggered conformation, which is shown in Fig. 12.5, these repulsions are at their lowest. Thus, during a complete revolution, the molecule passes three times through, first, a position of low potential energy (staggered conformation) and then, secondly, through three positions of high potential energy (eclipsed conformation).

The highest potential energy, V_{max}, which corresponds to an eclipsed conformation, is called the **barrier to internal rotation**. In ethane this barrier is not negligible and corresponds to a characteristic temperature for internal rotation, Θ_{ir}, of some 1380 K. Thus, although some excitation of internal rotation modes will be evident at room temperature, the internal rotation in ethane is by no means entirely free. On the other hand, the characteristic temperature for internal rotation of the methyl group in toluene is only 480 K, so this group will enjoy almost entirely free internal rotation at room temperature.

Torsional oscillation

The need to distinguish between internal rotation and torsional oscillation arises because the partition functions for these two types of motion are quite different. In **torsional oscillation**, the partition function is essentially one for a vibration. In this limit, the internal energy and heat capacity (as well as other thermodynamic quantities) will tend to values that are characteristic of **vibration**

$$q_{ir} \Rightarrow q_{vib} \qquad C_{V,ir} \Rightarrow R \tag{12.17}$$

Free rotation

In **internal rotation**, however, the partition function is essentially one for a rotation. In this limit, the internal energy and heat capacity (as well as other thermodynamic quantities) will tend to values that are characteristic of **rotation**

$$q_{ir} \Rightarrow q_{rot} \qquad C_{V,ir} \Rightarrow \tfrac{1}{2}R \tag{12.18}$$

Limiting and intermediate cases

In terms of the characteristic internal rotation temperature, θ_{ir}, it is a relatively simple matter to set up criteria for the two extremes of behaviour. If θ_{ir}/T is 10 or more, internal rotation becomes a rare event and **torsional oscillation** dominates. On the other hand, if θ_{ir}/T is 0.1 or less, free **internal rotation** dominates and torsional oscillation vanishes. In between these limits, a full calculation is needed. This can be done, but details lie outside the scope of this outline

Some values at room temperature of θ_{ir}/T for the methyl group in a number of different molecules are shown in Table 12.3. It is clear that certain methyl groups can spin quite freely (as in 1,2-dimethylethyne, methylbenzene, and propanone) whereas others (as in ethane, propane, tetramethylsilane, and dimethylpropane) tend much more towards pure torsional oscillation. In between, the problem becomes much less tractable, but it is one that raises particular interest nowadays as, for example, in the conformations and consequent thermodynamic properties available to large biological molecules.

In torsional oscillation and free internal rotation, the **reduced moment of inertia** of the rotating group, I_{red}, assumes the role normally played by the **reduced mass** in forming the appropriate partition function.

$$I_{red} = \frac{I_1 I_2}{I_1 + I_2} \tag{12.19}$$

where I_1 and I_2 are the moments of inertia of the rotating group and the rest of the molecule about the axis of internal rotation.

In free internal rotation, it is also necessary to introduce a **symmetry number for internal rotation**, σ_{int}, which denotes the number of indistinguishable configurations through which the rotating group passes in each 360° revolution. For the methyl group, for example, $\sigma_{int} = 3$.

Table 12.3 Relative internal rotation barriers at 300 K

Species	θ_{ir}/T
Me C≡C Me	0
⬡– Me	1.6
Me OH	1.6
Me_2CO	1.6
Me_4Si	2.1
Me CH=CH Me	3.3
Me CH=CH_2	3.5
Me OH–CH_3	4.6
Me CCl_3	5.0
Me CH_2 Me	5.9
Me CH_2Cl	6.2
Me_4C	7.5

Internal symmetry number, σ_{int}.

12.11 Conclusions

Vibrational energy levels are much more widely spaced than rotational energy levels. **Characteristic temperatures** for rotation take values that are of the order of tens or hundreds of kelvins; those for vibration are commonly in the region of hundreds or thousands of kelvins. Consequently, it is rare for many vibrational states above the ground state to be occupied. Frequently, one can assume the value of 1 for q_{vib} and avoid making more detailed calculations. This is especially true for all vibrational modes involving light atoms such as hydrogen, say.

Molecules consisting of many atoms can have large numbers $(3N - 6)$ of individual vibrational modes. **Normal mode analysis** allows each independent mode to be assigned and identified in the appropriate spectrum. Although the contribution to the vibrational partition function of each mode may be quite small individually, the sheer number of these modes in even moderately complex molecules may well make the total vibrational contribution quite appreciable.

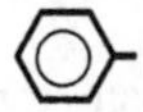

13 The electronic partition function

13.1 Introduction

Atoms and molecules can store energy in the form of excited electronic energy. They can acquire electronic excitation energy and rise in energy to states above the ground electronic state. Unlike excited states in translation, rotation, and vibration, for which a regular progression of energy states can be formulated, excited electronic states show no regular or predictable series of increasing energies in their excited states. Consequently, the sum over states which q_{el} represents does not lend itself to summation in a closed form. Nor can integration be invoked, not least because successive electronic energy states are not at all close and do not, therefore, constitute anything like a virtual continuum.

A simplified expression for q_{el}, other than the basic summation over successive electronic states, cannot therefore be found. Fortunately, this causes no problems in real systems since, with rare exceptions, one of which we shall consider below, *molecules* have excited electronic states that are very much higher in energy than their respective ground states. As we shall see, this enables us to truncate the "sum of states" at a very early stage, most often at its first term, so the irregularity of the succeeding series poses no further problems. Electronic partition functions cannot be ignored, but neither do they give rise to many difficulties.

13.2 The characteristic electronic temperature, θ_{el}

Typically, characteristic electronic temperatures, q_{el}, are of the order of several tens of thousands of kelvins. Such high excitation energies mean that excited electronic states in molecules remain unpopulated unless the temperature rises to several thousands of kelvins. Consequently, only the first (ground-state) term of the electronic partition function need ever be considered at temperatures in the range from ambient to moderately high.

It is tempting at this stage to decide that q_{el} will not be a significant factor in our investigations into the statistical behaviour of molecules. Once we have assigned to it, as is our custom, the energy $\varepsilon_0 = 0$, we might conclude that

$$q_{el} = \sum_i e^{-\varepsilon_{el,i}/kT} = e^{-0} + 0\,(\text{higher terms}) = 1 \qquad (13.1)$$

To do so, however, would be unwise, because eqn 13.1 ignores the possible **degeneracy** of the ground electronic. The effect of this on the electronic partition function is dealt with in the next section, and some illustrative data for halogen atoms are given at the end of Section 13.4.

13.3 Degeneracy

The correct expression to use in place of eqn 13.1 must include the degeneracies of the electronic states in the summation. thus

$$q_{\text{el}} = \sum_i g_i e^{-\varepsilon_{\text{el},i}/kT} = g_0 e^{-0} + 0\,(\text{higher terms}) = g_0 \qquad (13.2)$$

For O_2, $g_0 = 3$

Most molecules and stable ions have non-degenerate ground states. A notable exception to this generality is molecular oxygen, O_2, which has a ground-state degeneracy of three. Thus, for many stable ions and molecules, eqn 13.1 is adequate simply because $g_0 = 1$.

Atoms, however, frequently have ground states that are degenerate. The degeneracy of electronic states is determined by the value of the **total angular momentum quantum number,** J, which appears as the right subscript of the **term symbol** representing the electronic state in question. Taking the symbol Γ as a general term in the **Russell–Saunders spin-orbit coupling** approximation, we denote the spectroscopic state of the ground state of an **atom** as:

$$\text{Spectroscopic atomic ground state} = {}^{(2S+1)}\Gamma_J \qquad (13.3)$$

where S is the **total spin angular momentum quantum number** which gives rise to the **term multiplicity** $(2S+1)$. The degeneracy, g_0, of the electronic ground state in **atoms** is related to J through

$$g_0 = 2J+1 \quad (\text{atoms}) \qquad (13.4)$$

A full explanation of these and other features of atomic spectra may be found in Softley (1994).

For diatomic **molecules,** the term symbols for electronic states are made up in much the same way as those for atoms. What is important now, however, is the component of the **total orbital angular momentum** about the **internuclear axis.** This determines the term symbol used for the molecule (Σ, Π, Δ, etc, corresponding to S, P, D, etc. in atoms). As with atoms, the term multiplicity $(2S+1)$ is added as a superscript to denote the multiplicity of the molecular term but, in the case of molecules, it is this term multiplicity that represents, uniquely, the degeneracy of the electronic state. Thus, for diatomic **molecules,** we have

$$\text{Spectroscopic molecular ground state} = {}^{(2S+1)}\Gamma \qquad (13.5)$$

for which the ground-state degeneracy is

$$g_0 = 2S+1 \quad (\text{molecules}) \qquad (13.6)$$

Excited electronic-state degeneracies for atoms and for diatomic molecules can be deduced in a similar manner. Table 13.1 shows some atomic and

Table 13.1 Degeneracies of electronic states of atoms and simple molecules. Where there are low-lying excited states, characteristic electronic temperatures, θ_{el}, are also shown

Species	Term Symbol	g_n	θ_{el}/K
Li	${}^2S_{1/2}$	g_0=2	
C	${}^3P_0$	g_0=1	
N	${}^4S_{3/2}$	g_0=4	
O	${}^3P_2$	g_0=5	
F	${}^2P_{3/2}$	g_0=4	
	${}^2P_{1/2}$	g_1=2	590
Cl	${}^2P_{3/2}$	g_0=4	
	${}^2P_{1/2}$	g_1=2	1300
Br	${}^2P_{3/2}$	g_0=4	
	${}^2P_{1/2}$	g_1=2	5440
I	${}^2P_{3/2}$	g_0=4	
	${}^2P_{1/2}$	g_1=2	11 100
N_2	${}^1\Sigma_g^+$	g_0=1	
NO	${}^2\Pi_{1/2}$	g_0=2	
	${}^2\Pi_{3/2}$	g_1=2	178
O_2	${}^3\Sigma_g^-$	g_0=3	
	${}^1\Delta_g$	g_1=1	11 650
F_2	${}^1\Sigma_g^+$	g_0=1	

molecular electronic state degeneracies together with some energy gaps (θ_{el}), for the energy gap between the ground electronic state and the low-lying first excited state, where these are low enough to be relevant.

13.4 The electronic partition function

In the vast majority of case, where the energy gap between the ground and the first excited electronic state is large, the electronic partition function simply takes the value g_0, as in eqn 13.2.

However, when the ground-state to first excited state gap is **not** negligible compared with kT (θ_{el}/T is not very much less than unity), it is necessary to consider the first excited electronic state which, under these circumstances, may be appreciably populated. There is, happily, rarely any need to consider higher excited electronic states because these have much higher energies. If the first excited state is at all populated, however, the expression for the electronic partition function becomes

$$q_{el} = g_0 + g_1 e^{-\theta_{el}/T} \tag{13.7}$$

where θ_{el} is the characteristic temperature for the gap between the ground and the first excited electronic state.

For fluorine atoms, we calculate q_{el} at 1000 K as follows, using data from Table 13.1.

$$q_{el} = 4 + 2e^{-590/1000} = 5.109$$

An example of the use of eqn 13.7 is shown alongside for fluorine atoms at 1000 K and further results for the other halogens are given in Table 13.2 which shows how the electronic partition function declines towards its minimum value ($g_0 = 4$) as the gap between the ground and first excited electronic states widens until, for iodine atoms, only the electronic ground state is populated. Note that in this case the lowest value of the partition function is *not* 1 but 4, because the ground-state electronic degeneracy is 4.

As was asserted in Section 13.2, the postulate that the lowest value of q is always unity, is incorrect. The lowest value of any factor in the molecular partition function is its ground state degeneracy. For translation, rotation, and vibration, $g_0 = 1$ so the minimum value of the partition function for these modes is also 1; for electronic energies this need not always be so. Table 13.2 shows q_{el} for each of the four halogen atoms at 1000 K. The lowest value possible would be 4, which is the degeneracy of the ground electronic state, and the maximum value possible is 6, when both ground and first excited states are fully occupied. So, in fluorine the first excited state is just over half-populated, in chlorine about one-quarter populated, and in bromine and iodine hardly populated at all.

Table 13.2 The electronic partition function for halogen atoms at 1000 K

Halogen atom	q_{el}
F	5.109
Cl	4.545
Br	4.009
I	4.000

13.5 The singular case of NO

In Table 13.1, nitrogen monoxide, NO, and the very narrow gap between the ground and first excited electronic states of this molecule, is noticeable. This narrow gap suggests that the first excited state in NO can become populated at quite low temperatures. Because the gap to the next excited electronic state (the second) is very large, the ground and first excited states in NO provide a singular example of an **isolated two-state system**. This case was not used as an example in Chapter 6 because it would not have been as straightforward then to provide an explanation for the origin of the two isolated states.

In the nitrogen monoxide molecule, both the orbital and spin angular momenta can take either of two orientations relative to the molecular axis. So, there are four possible states, two of these comprise a lower ground-state doublet electronic state and the other two of which form a low-lying doublet first excited electronic state only 178 K (in characteristic temperature terms) above this ground state. So we have

$$g_0 = g_1 = 2; \qquad \theta_{el} = 178\text{ K}; \qquad \therefore q_{el} = 2 + 2e^{-178/T} \tag{13.8}$$

From this molecular partition function for electronic states, it is a simple matter to derive an expression for the electronic heat capacity. Since the ground and first excited states are both degenerate, the most appropriate expression is eqn 6.16. However, since the degeneracies of the two states are identical ($g_1/g_0 = 1$), an equally relevant heat capacity expression is eqn 6.13

$$C_V = Nk(\beta\Delta\varepsilon)^2 \left[\frac{e^{\beta\varepsilon\Delta}}{(e^{\beta\varepsilon\Delta} + 1)^2} \right] \tag{6.13}$$

which, in molar (and characteristic electronic temperature) terms, gives

$$\frac{C_{el,m}}{R} = \left(\frac{\theta_{el}}{T}\right)^2 \left[\frac{e^{\theta_{el}/T}}{\left(e^{\theta_{el}/T} + 1\right)^2} \right] \tag{13.9}$$

Equation 13.9, with θ_{el} = 178 K, is plotted in Fig. 13.1. This plot provides an excellent representation of the experimental data of Eucken and d'Or (1932). The classical heat capacity of 2.5 R (1.5 R for translation plus 1.0 R for a linear rotor) has been subtracted from the experimental results to leave only the anomalous electronic contribution. The maximum additional electronic heat capacity is close to 0.44 R (as predicted by the two-state model in Ch. 6) at a temperature of 74 K (very close to the theoretical prediction of 0.42 × 178 K = 74.8 K for a two-state system).

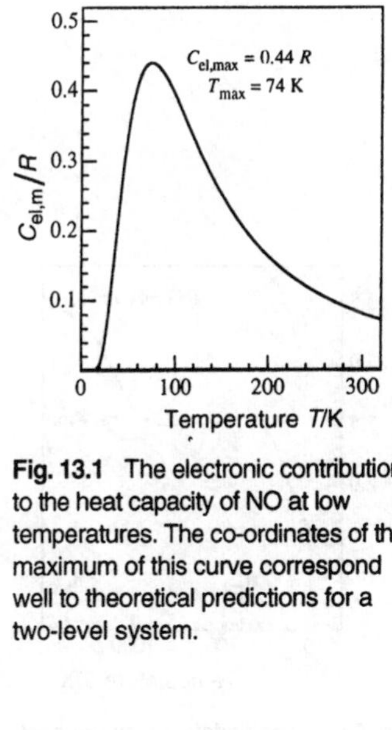

Fig. 13.1 The electronic contribution to the heat capacity of NO at low temperatures. The co-ordinates of the maximum of this curve correspond well to theoretical predictions for a two-level system.

13.6 Conclusions

Electronic states enter into the partition function almost always only through the degeneracy of the ground electronic state. It is rare to find an electronic first excited state that is close enough to the ground state to enjoy any sensible level of population at any but the most elevated temperatures. If there is a low-lying first excited state, as in nitrogen monoxide, excellent and very direct confirmation of predicted two-state behaviour is obtained.

References

Eucken, A. and d'Or, L. (1932). *Nachr. Akad. Wiss. Götingen, Math. Physik Kl.*, 107.

Softley, T. P. (1994). *Atomic Spectra*. OCP 19. OUP, Oxford.

14 Heat capacity and Third Law entropy: two case studies

14.1 Introduction

Having exhausted our survey of all of the energy modes that contribute to the partition function, we can now set about the business of applying the knowledge we have gained. In this chapter, we tackle two practical questions.

- What is the temperature dependence of the overall heat capacity of a molecule which has internal modes?
- How well does the overall entropy, as calculated from the partition function, correspond to calorimetric measurements? Is the agreement perfect, or merely quite good?

14.2 The heat capacity $C_{V,m}$ as a function of temperature

The overall heat capacity is simply the sum of the individual contributions from each energy mode. When the mode in question is "classical", i.e. at equipartition, when the external temperature is much greater than the characteristic temperature for that mode ($T \gg \theta$) and quantum effects play no apparent part; its contribution to the heat capacity remains constant with temperature. Conversely, when the external temperature is much lower than the characteristic temperature for that mode ($T \ll \theta$), the mode remains unexcited and its contribution to the heat capacity is zero. In between, the heat capacity varies with temperature.

Some of these modes (like translational modes) are classical at all temperatures; others (like rotational modes) become classical at quite low temperatures. For others still (like vibrational modes) the external temperature rarely gets high enough for the mode to do much more than start to become excited and to contribute just a little to the overall heat capacity.

Fig. 14.1 The variation of molar heat capacity with temperature in hydrogen deuteride. Rotational or vibrational contributions become significant only close to and above their characteristic temperatures. The maximum in the rotational contribution is discussed in Section 14.3.

The general picture, then is one of successive steps and plateaux for $C_{V,m}(T)$ as successive modes come into play. This is illustrated in Fig. 14.1 for hydrogen deuteride which has $\theta_r = 64.1$ K and $\theta_{vib} = 5300$ K. As has been our custom, the temperature is scaled logarithmically in order to reveal more clearly the behaviour in different temperature regions.

The heat capacity starts from $\frac{3}{2}R$ at the lowest temperatures when only the three translational modes, which are always active, contribute to the heat capacity. As the temperature rises towards and then past θ_r, the two rotational modes become active and, after rising to a maximum of $C_{rot,m} = 1.1\,R$ (explained below) at a temperature of $0.8\,\theta_r$, eventually decline gradually to their equipartition share of $1.0\,R$, raising the overall heat capacity to $\frac{5}{2}R$.

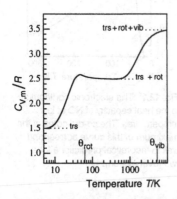

A further (but much larger) rise in the temperature brings the single vibrational mode into play, eventually contributing a further amount, R, once the temperature gets well above θ_{vib}. At this point the total (and maximal) heat capacity is $\frac{7}{2}R$.

In reality, neither the low temperature loss of the rotational modes as the temperature is lowered, nor the high temperature gain of the vibrational modes are at all easy to observe experimentally in other molecules. For rotation this is always so, since the rotational temperature for most molecules is never very much above the absolute zero, and decreases rapidly with increasing molecular complexity (via the increased mass). For vibration in diatomic molecules, the vibrational temperature is usually much too high (several thousand kelvins) to allow measurement of finite vibrational heat capacities. However, some vibrational modes in more complicated (and more massive) molecules have characteristic temperatures of the order of a few hundred kelvins, and the rise to an equipartition value for such modes can often be observed experimentally.

14.3 The maximum in $C_{rot,m}$

The slight but significant maximum in $C_{rot,m}$ for HD ($1.1\,R$ at $0.8\,\theta_r$) can be predicted. It arises for precisely the same reasons as the maxima we observed for two-level systems (Chapter 6) and for low-lying electronic states (Chapter 13). At low temperatures, low, that is, with respect to the energy level spacing, only the ground state ($J = 0$) in rotation is occupied. As the temperature increases, molecules begin to acquire enough rotational energy to start to populate the first excited rotational state ($J = 1$).

At this juncture, the rotational system behaves just like an **isolated two-state system**, with particles now able to populate the upper state through thermal excitation. The natural progression for a two-state system (see Section 6.5) would be to have $C_{V,m}(T)$ rise to a maximum before declining to zero as the upper level becomes saturated.

For rotation, as the temperature rises, the rotational system begins to establish communication between the ground and the first excited states but has, as yet, no means of sensing the second excited state lying even higher above the first. The population and energy shifts bring $C_{rot,m}$ towards a maximum but, unlike the isolated two-state case, communication with the third and then higher levels leads the system inexorably towards equipartition.

The maximum in the $C_{rot,m}$ curve results from the two lowest levels becoming isolated from higher levels at low enough temperatures; its rather unimpressive size comes about because this isolation is broken as the temperature rises and molecules start to flow from ground to first excited to second state, and so on.

Working to enhance this tendency are the degeneracies of successive rotational states. The ratio, $g(n+1)/g(n)$, of successive level degeneracies is shown in Table 14.1. Largest by nearly a factor of two of these is the g_1/g_0 ratio. In Chapter 6 (Fig. 6.5), we saw that as the degeneracy ratio becomes larger, the peak in the $C_{V,m}(T)$ graph becomes higher and shifts to slightly lower temperatures. In this case, the major effect is seen through the first

Table 14.1 The degeneracies of low-lying rotational levels and their ratios.

rotational level n	degeneracy g_n	ratio g_{n+1}/g_n
0	1	3.00
1	3	1.67
2	5	1.40
3	7	1.28
4	9	1.22

excited to ground state degeneracy ratio, which leads to a relatively sudden increase in $C_{rot,m}$ at a temperature just below θ_r, when large numbers of molecules begin to become excited to the $J = 1$ state. After this sudden rush, there is no greatly enhanced tendency for early excitation of molecules into higher states, so the heat capacity falls gradually to its classical, equipartition, value.

14.4 Calorimetric and spectroscopic entropy

Calorimetric (or Third Law) entropy, S_{cal}

The entropy change, ΔS, that occurs when a substance at constant pressure is heated from temperature T_1 to temperature T_2 is given by either of the definite integrals

$$\Delta S = \int_{T_1}^{T_2} \frac{C_p}{T}\,dT \quad \text{or} \quad \Delta S = \int_{\ln T_1}^{\ln T_2} C_p\,d\ln T \qquad (14.1)$$

which can be determined by measuring the area underneath the appropriate heat capacity–temperature plot.

Phase transitions occur with changes of state (s→l and l→g) but may also, and less detectably, occur with phase changes in the solid state.

Equation 14.1 is incomplete and also needs to take account of any changes in entropy at constant temperature, such as those that take place within the range of integration at any **phase transition**. We need, therefore, to add to the right-hand side of eqn 14.1, an amount $\Delta H_{tr}/T_{tr}$ for each such transition. ΔH_{tr} is the enthalpy change of the phase transition and T_{tr} is the temperature at which it occurs. To account for all phase transitions within the range of integration, we write

$$\Delta S = \int_{T_1}^{T_2} \frac{C_p}{T}\,dT + \sum_i \left(\frac{\Delta H_{tr}}{T_{tr}}\right)_i \qquad (14.1a)$$

where i denotes the ith phase transition in the interval T_1 to T_2

$S_{T_2} - S_0 = S_{cal}$

$S_0 = 0$ (for perfectly crystalline substances)

If the lower temperature limit is chosen as 0 K, the integral determines the value of the quantity $S_{T_2} - S_0$, the difference between the entropy at temperature T_2 and its value at the absolute zero. The **Third Law of thermodynamics** sets this latter value, S_0, at zero for perfect crystals and, in fact, the procedure used to evaluate the integral in eqn 14.1a involves graphical extrapolation from the lowest experimental temperature back to the origin of the graph.

It is the value of the entropy, determined using the Third Law in conjunction with eqn 14a, that is known as the *calorimetric* or *Third Law* entropy, S_{cal}. The label **calorimetric** arises from the fact that the bulk of the data that lead to a value for eqn 14.1a are determined from measurements of heat changes established using calorimetry.

The defining expressions for calorimetric/Third Law entropy are shown in eqn 14.1b.

$$S_{cal} = \int_0^{T_2} \frac{C_p}{T}\, dT + \sum_i \left(\frac{\Delta H_{tr}}{T_{tr}}\right)_i$$

(14.1b)

$$= \int_{-\infty}^{\ln T_2} C_p \, d\ln T + \sum_i \left(\frac{\Delta H_{tr}}{T_{tr}}\right)_i$$

Spectroscopic entropy, S_{spec}

The spectroscopic entropy is so named because it is determined via the partition function using energy level spacings; a knowledge of these spacings can be obtained from spectroscopic measurements. The spectroscopic entropy is defined by eqn 14.2, with the running variable i in the summation denoting different energy modes, such as translation, rotation, vibration, electronic, and the like, and each of the S_i derived from the toolkit function for entropy and the appropriate canonical partition function, Q_i

$$S_{spec} = \sum_i S_i$$

with (14.2)

$$S_i = k \ln Q_i + kT\left(\frac{\partial \ln Q_i}{\partial T}\right)_V$$

Toolkit function

$$S = k\ln Q + kT\left(\frac{\partial \ln Q}{\partial T}\right)_V$$

Discrepancies

The difference between the statistical mechanical **spectroscopic** entropy, S_{spec}, and the experimental **calorimetric** entropy, S_{spec}, is called the **residual** entropy, S_{resid}.

$$S_{resid} = S_{spec} - S_{cal} \quad (14.3)$$

When there are no complications, the expectation that calorimetric and spectroscopic entropy are identical is borne out rather well. For example, the calorimetric and spectroscopic entropies of carbonyl sulfide, OCS, can be compared, as in Table 14.2. The calorimetric determination gives a value that is very close to the spectroscopic value. Calorimetry and spectroscopy agree well, and $S_{resid} = 0$, as Table 14.2 illustrates: when there are no complications! The complications that can arise come about for a variety of reasons. In the sections that follow, we explore those that are most important.

Table. 14.2 The calorimetric and spectroscopic entropy of OCS at 298.15K.

S_{cal}/J K^{-1} mol^{-1}	S_{spec}/J K^{-1} mol^{-1}
231.24	231.54

14.5 Residual entropy

Residual entropy is defined in eqn 14.3. A selection of reasons why the residual entropy $S_{resid} \neq 0$ is explored below. In general, it is usual to question the experimental results before seeking to cast doubt on the validity of the statistical mechanical outcome.

In general, there are three possible reasons for a value of $S_{resid} \neq 0$.

- An undetected low-temperature phase transition involving an unexpected solid-state phase change and an entropy $\Delta S = \Delta H/T$.
- Use of the wrong low temperature degeneracy.
- Some remaining disorder in the crystal, even at the absolute zero, sow that the **Third Law** premise ($S_0 = 0$) is invalid. In other words, $S_0 > 0$.

While the first two of these cannot ever be discounted without further investigation, they differ from the third in that it is essentially our own fault that they appear to point to a discrepancy. There is no discrepancy; we are simply not smart enough to realise this!

The third reason is real. For the Third Law states:

There is no contribution to the entropy at the absolute zero from any source in internal equilibrium.

A frozen-in, non-zero entropy, from whatever source, can continue to contribute to the entropy at the absolute zero, so that $S_0 > 0$.

Frozen-in random orientations, (i) CO: $S_{resid} = 0.6R$

Ordered lattice	CO CO CO CO CO CO CO CO CO CO CO CO
Another ordered lattice	CO CO CO CO OC OC OC OC CO OC CO OC
Disordered lattice	CO CO OC CO OC CO OC OC OC OC CO OC

The crystal lattice of CO at the absolute zero should be perfectly ordered. However, the CO molecule has such a low dipole moment that the two possible orientations in the crystal lattice, OC and CO, are virtually equivalent in energy. The low dipole exerts only a small ordering effect. As the temperature drops, the energy difference between the two orientations, $\Delta\varepsilon = \varepsilon(\text{CO–CO}) -- \varepsilon(\text{CO–OC})$, becomes comparable to kT only below the temperature at which CO freezes. At this point, the activation energy required to convert the configuration CO to the configuration OC makes the re-orientation very slow. Consequently, any real CO crystal close to the absolute zero is not perfectly ordered, $S_0 > 0$, and there is an observed residual entropy of about 5 J K^{-1} mol^{-1}.

Were the lattice totally disordered, we could calculate the entropy of disorder as follows: there are N lattice sites and 2 possible orientations. This gives

$$\therefore\ S_{resid} = k \ln \Omega, \quad \text{with } \Omega = 2^N \text{ ways}$$

$$\text{so } S_{resid} = k \ln 2^N = Nk \ln 2$$

$$\text{and, with } Nk = R,\ S_{resid} = R \ln 2 = 5.8 \text{ J K}^{-1} \text{ mol}^{-1}$$

$$\text{as against} \quad S_{resid} = S_{spec} - S_{cal} = 4.6 \text{ J K}^{-1} \text{ mol}^{-1}$$

The difference here can be accounted for by suggesting some (slight) dipole ordering producing a less than random lattice. Essentially, the agreement is considered to be adequate.

Frozen-in random orientations, (ii) NO: $S_{resid} = 0.35R$

Nitrogen monoxide, already the subject of a low temperature increase in heat capacity on account of its singular electronic structure with an unpaired electron, tends to dimerise at low temperatures (by combining the unpaired spins) to form the rectangular planar molecule N_2O_2.

$$2NO \rightleftharpoons N_2O_2$$

Again, the random arrangements of these dimers produces an entropy of $R \ln 2$ **per mole of dimer**. But the low temperature crystal consists of half as many molecules of N_2O_2 as there were molecules of NO originally, so

N_2O_2 has two possible orientations

N - O O - N
I I I I
O - N N - O

$$\therefore\ S_{resid} = k \ln \Omega, \text{ per mole of } N_2O_2$$
$$\text{so } S_{resid} = \tfrac{1}{2} k \ln 2^N = \tfrac{1}{2} Nk \ln 2 \text{ per mole of NO}$$
$$\text{and, with } Nk = R,\ S_{resid} = \tfrac{1}{2} R \ln 2 = 2.88 \text{ J K}^{-1} \text{ mol}^{-1}$$
$$\text{as against} \quad S_{resid} = S_{spec} - S_{cal} = 2.89 \text{ J K}^{-1} \text{ mol}^{-1}$$

The agreement here, between theory and experiment, is excellent.

Frozen-in random orientations, (iii) CH_3D: S_{resid} = 1.39*R*

There are four equivalent tetrahedral positions that the deuterium atom in solid CH_3D can occupy. With four equivalent orientations of approximately equal energy, the residual molar entropy will be

$$\therefore\ S_{resid} = k \ln \Omega, \text{ with } \Omega = 4^N \text{ ways}$$
$$\text{so } S_{resid} = k \ln 4^N = Nk \ln 4$$
$$\text{and, with } Nk = R,\ S_{resid} = R \ln 4 = 11.52 \text{ J K}^{-1} \text{ mol}^{-1}$$
$$\text{as against} \quad S_{resid} = S_{spec} - S_{cal} = 11.56 \text{ J K}^{-1} \text{ mol}^{-1}$$

Once again, the agreement between theory and experiment is excellent. Equally good agreement is found for $FClO_3$, where the fluorine atom can occupy one of four energetically equivalent tetrahedral positions in the crystal lattice.

Frozen-in random orientations (iv), ice-I: S_{resid} = 0.41*R*

Ice-I is the crystal form adopted by water in the solid state at atmospheric pressure. It is just one of the several crystalline forms adopted by water.

The solid structure contains covalent σ bonds and hydrogen bonds, two of each attached to the same oxygen atom per molecule of H_2O. Each water molecule can orient its σ-bonded hydrogens in six different ways (a tetrahedron has six edges). However, the probability that any one chosen direction is available to (or free to accommodate) a given hydrogen is only 1 in 2, since each neighbouring water molecule has only half of its tetrahedral sites available: two of the four positions are occupied by σ-bonded hydrogens, the other two are vacant.

The probability that a second chosen direction should be vacant is 1 in 2, also, giving the overall probability for fixing two chosen directions as $(\tfrac{1}{2})^2$, or a 1 in 4 chance. So, for each water molecule, there are 6/4 = 3/2 choices and, for N water molecules there is a total of $(3/2)^N$ possible orientational arrangements.

Thus, for the entropy arising from these random orientations, we have

$$\therefore\ S_{resid} = k \ln \Omega, \quad \text{with } \Omega = \left(\frac{3}{2}\right)^N$$

$$\text{so } S_{resid} = k \ln (1.5)^N = Nk \ln (1.5)$$

$$\text{and, with } Nk = R,\ S_{resid} = R \ln (1.5) = 3.37 \text{ J K}^{-1} \text{ mol}^{-1}$$

$$\text{as against} \quad S_{resid} = S_{spec} - S_{cal} = 3.4 \text{ J K}^{-1} \text{ mol}^{-1}$$

As with the previous cases, the agreement between theory and experiment is excellent.

14.6 Conclusions

These two case studies have allowed us to use statistical thermodynamics to shed light on two potentially puzzling phenomena. These have involved gaining an understanding of the temperature variation of the heat capacity of simple diatomic molecules and more complex polyatomic species, behaviour that was inexplicable before the advent of quantum theory and the concept of quantised energy states.

In the process, we have also found an explanation for some quite subtle features, such as the low maximum in the (C_V,T) curve close to the characteristic rotational temperature, θ_r, which can be understood using the concepts developed in Chapter 6.

The use of statistical thermodynamics to provide a theoretical value for absolute entropies has also proved to be a success. The fact that entropies can be calculated with such certainty is already a major reward for our endeavours. That we can unravel such subtleties as the contributions to the entropy of random orientations among molecules makes this reward that much more significant.

15 Calculating equilibrium constants

15.1 Introduction

In this final chapter we reach the point at which it becomes possible to apply earlier ideas to the problem of calculating an equilibrium constant from knowledge of the molecular mass, electronic configuration, geometry, and bond characteristics of each of the species that take part in the equilibrium. Much of this basic information can be obtained from the various spectra characteristic of the participating species, so it is to the spectral data that we shall turn to find the information we need.

As in Chapter 7, the development of the required formalism is not particularly complicated but, unlike the case of the derivation of a thermodynamic toolkit, there are insights to be gained from the early sections of this chapter, and especially Section 15.4, that had no counterpart in Chapter 7.

Deriving the relationship between partition functions and equilibrium constants brings a greater understanding of the true nature of chemical equilibrium and is worth doing for its own sake. But being able to use the resulting equations to give quantitative results is also an invaluable skill, hence we shall end this chapter, and the whole book, by applying our understanding to some real situations, and by calculating values of actual equilibrium constants.

15.2 The molar Gibbs free energy

For perfect gases, we can express the molar Gibbs function in terms of the molar partition function q_m, using eqn 7.29 (Section 7.11)

$$G_m - G_m(0) = -RT \ln\left(\frac{q_m}{L}\right) \qquad (7.29)$$

The equilibrium constant, $K_p^{\ominus}$, is related to the change in the standard molar Gibbs function, $\Delta G^{\ominus}$

$$\Delta G^{\ominus} = RT \ln K_p^{\ominus} \qquad (15.1)$$

Equation 7.29 can be combined with eqn 15.1 to yield an expression for the equilibrium constant, K. First, we express eqn 7.29 with all thermodynamic variables referred to their standard states

$$G^{\ominus}_{\mathrm{m}} = G^{\ominus}_{\mathrm{m}}(0) - RT\ln\left(\frac{q^{\ominus}_{\mathrm{m}}}{L}\right) \qquad (15.2)$$

$p^{\ominus} = 1$ bar

For 1 mole of gas

$$q_{\mathrm{trs}} = \left[\frac{2\pi m}{h^2\beta}\right]^{\frac{3}{2}} V_{\mathrm{m}}$$

The **standard molar partition function,** $q^{\ominus}_{\mathrm{m}}$, is the value of the molar partition function under standard conditions, when $p = p^{\ominus}$. Now, only the translational partition function is dependent on the volume (and hence, through the equation of state, on the pressure), so we evaluate $q^{\ominus}_{\mathrm{m}}$, by replacing the molar volume, V_{m}, by the standard state volume, $V^{\ominus}_{\mathrm{m}} = RT/p^{\ominus}$.

From eqn 15.2, and writing the molar partition function for component i as $q_{\mathrm{m},i}$, we can denote the standard molar Gibbs free energy change, $\Delta G^{\ominus}_{\mathrm{m}}$, for the reaction as

$$\Delta G^{\ominus}_{\mathrm{m}} = \sum_i \nu_i G^{\ominus}_{i,\mathrm{m}} = \sum_i \nu_i G^{\ominus}_{i,\mathrm{m}}(0) - RT\sum_i \nu_i \ln\left(\frac{q^{\ominus}_{i,\mathrm{m}}}{L}\right) \qquad (15.3)$$

where the ν_i are the stoichiometric coefficients of the balanced equilibrium equation.

For a perfect gas at absolute zero, $G(0) = U(0)$, a result that follows from eqn 7.22a

$$U(0) = H(0) = A(0) = G(0) \qquad (7.22a)$$

so that

$$\sum_i \nu_i G^{\ominus}_{i,\mathrm{m}}(0) = \Delta G^{\ominus}_{\mathrm{m}}(0) = \Delta U^{\ominus}_{\mathrm{m}}(0) \qquad (15.4)$$

The only form of energy that does not drop to a common value (conventionally taken as zero) at absolute zero is the vibrational energy. At absolute zero, vibrational energy falls to the zero-point energy, $\frac{1}{2}h\nu_0$. This quantity varies from species to species. In Section 12.7, we established that, for each component, the molar zero-point energy is given by (eqn 12.11)

$$U(0)_{\mathrm{vib,m}} = \frac{1}{2}Lh\nu_0 \qquad (15.5)$$

$\Delta U^{\ominus}(0)$ is simply the difference between the molar zero-point energies (*zpe*) of products and reactants, each component weighted by its appropriate stoichiometric coefficient, ν, with ν_i*(products)* positive and ν_j*(reactants)* negative.

$$\Delta U^{\ominus}(0) = \sum_{products,\, i} \nu_i\, zpe_i - \sum_{reactants,\, j} \nu_j\, zpe_j$$

By convention, the change in standard internal energy at the absolute zero, $\Delta U^{\ominus}(0)$, is given the symbol ΔE_0 per mole, or $\Delta\varepsilon_0$ per molecule. Thus, we can write

$$\Delta E_0 = \Delta U^{\ominus}(0) = \sum \nu_i U^{\ominus}_i(0) \qquad (15.6)$$

15.3 The equilibrium constant, $K_p^{\ominus}$

We are now in a position to re-write eqn 15.3 in a way that will lead to an expression for the equilibrium constant. Firstly, recognising that the sum of logarithmic terms is the same as the logarithm of the product of these terms,

$\ln a_1 + \ln a_2 + \ln a_3 \ldots = \ln(a_1 a_2 a_3)$

$$\sum_i \nu_i \ln\left(\frac{q_{i,\mathrm{m}}^{\ominus}}{L}\right) = \ln \prod_i \left(\frac{q_{i,\mathrm{m}}^{\ominus}}{L}\right)^{\nu_i} \qquad (15.7)$$

Π denotes *the product of*, just as Σ denotes *the sum of*.

Next, we replace $\sum \nu_i G_{i,\mathrm{m}}^{\ominus}(0)$ by ΔE_0, which gives the overall equation

$$\Delta G_{\mathrm{m}}^{\ominus} = -RT\left[-\frac{\Delta E_0}{RT} + \ln \prod_i \left(\frac{q_{i,\mathrm{m}}^{\ominus}}{L}\right)^{\nu_i}\right] \qquad (15.8)$$

an equation cast in this form to make it simple to compare with eqn 15.1 and to recognise that

$$\ln K = -\frac{\Delta E_0}{RT} + \ln \prod_i \left(\frac{q_{i,\mathrm{m}}^{\ominus}}{L}\right)^{\nu_i} \qquad (15.9)$$

whence we have

$$K = \prod_i \left(\frac{q_{i,\mathrm{m}}^{\ominus}}{L}\right)^{\nu_i} \mathrm{e}^{-\Delta E_0 / RT} \qquad (15.10)$$

This is a most significant result. It will enable us to calculate equilibrium constants, now that we know how to evaluate both the partition functions and the energy ΔE_0.

15.4 Interpreting the equilibrium constant

Being able to calculate the value of an equilibrium constant, though a worthy end in itself, has an equal partner in the light that can be shed on the nature of the equilibrium constant and how external factors cause it to change. In this section we shall show how the statistical formulation provided by eqns 15.9 and 15.10 can shed light on the very nature of equilibrium.

The partition function concerns itself with energy levels and how molecules are distributed over those levels. The spacings of, and the differences between, energy levels are clearly in the realm of **energetics**; this is one strand in our understanding of equilibrium. The way in which molecules spread themselves over the available energy levels, on the other hand, is clearly in the realm of **entropy**; the more accessible energy states are, the more ways molecules can spread to occupy them and the greater the entropy will be; this is the other strand in our understanding of equilibrium.

Classical thermodynamics, of course, already tells us this, since, when we combine eqn 15.1

$$\Delta G^{\ominus} = RT \ln K_p^{\ominus} \qquad (15.1)$$

with the relationship

$$\Delta G^{\ominus} = \Delta H^{\ominus} - T\Delta S^{\ominus}$$

the expression that results links the equilibrium constant to standard enthalpy and entropy changes

Equation 15.9:

$$\ln K = -\frac{\Delta E_0}{RT} + \frac{1}{R}\left[R\ln\prod_i\left(\frac{q^{\ominus}_{i,\mathrm{m}}}{L}\right)^{\nu_i}\right]$$

$$\ln K = -\frac{\Delta H^{\ominus}}{RT} + \frac{\Delta S^{\ominus}}{R} \qquad (15.11)$$

Equations 15.11 and 15.9 show clear similarities. In both cases the first, temperature-dependent, term is concerned with the energetics of the equilibrium; the actual enthalpy change for the reaction or the internal energy change at 0 K. Because of the inverse relationship with temperature, these are the terms that will dominate at sufficiently low temperatures; if the temperature is high enough (and here *high* is in relation to a characteristic temperature for the equilibrium of $\Delta H/R$ or $\Delta E_0/R$) these terms will eventually become negligible, leaving only the temperature-independent, entropic terms.

Equally, the final, temperature-independent terms have to do with entropy and bring into focus the way in which the partition function encapsulates entropy changes. The classical expression, derived without any reference to the structure or the quantised energy levels of molecules in a working fluid, helps us to give a physical interpretation to the origins of molecular entropy.

In the section that follows, we shall look at the different ways in which the enthalpy change (and ΔE_0) and the entropy change (and the partition function) influence the position of chemical equilibrium.

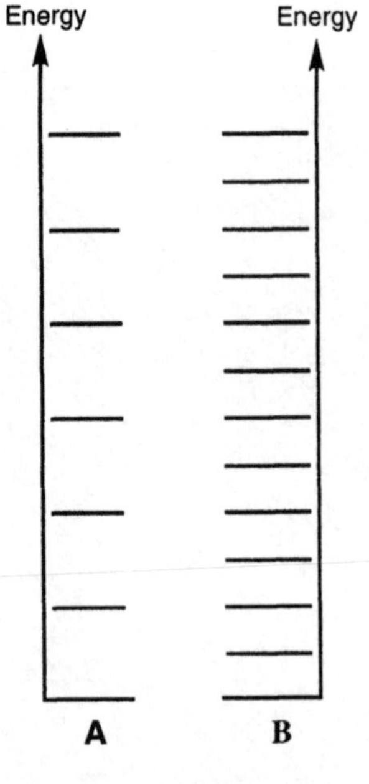

Fig. 15.1 Energy manifolds for a thermoneutral equilibrium. The 2:1 energy spacing dictates that the equilibrium constant $K \to 2$ **in the high temperature limit.**

15.5 Aspects of the equilibrium constant

Below, we explore the way in which the balance of thermal energy in an equilibrium process interacts with the entropy balance to determine the magnitude of the equilibrium constant in a dynamic system. We start with the simplest case, a thermoneutral process, in order to explore the major features influenced by the entropy and the partition function. In order to prevent the explanation from becoming too cluttered, we shall choose as our example a simple isomerisation reaction with unit stoichiometry

$$\mathrm{A(g)} \rightleftharpoons \mathrm{B(g)} \qquad (15.12)$$

Thermoneutral equilibrium

Imagine that the isomerisation in eqn 15.12 involves zero change in energy, so that $\Delta H^{\ominus} = \Delta E_0 = 0$. Under these circumstances, we can immediately equate the entropy change for the reaction to the logarithm of the ratio of the partition functions of A and B (eqns 15.9 and 15.11)

$$\Delta S^{\ominus} = R\ln\left(\frac{q^{\ominus}_{\mathrm{B,m}}}{q^{\ominus}_{\mathrm{A,m}}}\right) \qquad (15.13)$$

It this case, it is this ratio of the two partition functions that determines the position of equilibrium. The equilibrium will favour the side of the reaction that has the higher entropy, which, in terms of the partition functions, is the side of the reaction that has more energy states that are accessible.

ΔS is directly related to the logarithm of the ratio $\frac{q_B}{q_A}$.

Since, in the case of a thermoneutral reaction, there is no energetic bias one way or the other between A and B, the side of the equilibrium that will be favoured is the side with states spaced more closely, i.e. the side with a greater **density of states**. In Fig. 15.1, it is B that has the greater density of states (by a factor of 2), so the equilibrium will lie on that side.

Density of states, $g(\varepsilon)$: the number of energy states in unit energy interval $g(\varepsilon)=\frac{dN(\varepsilon)}{d\varepsilon}$

In principle, if one side has a density of states that increases more with increasing energy than the other, there is some scope for K to alter with temperature, though no obvious example springs to mind. Under such circumstances, the energy states of the side with the increasing density of states function will become more accessible as the temperature increases.

Thus, the ratio q_B/q_A is the determining factor in thermoneutral equilibrium. The situation, in classical terms, is one where $\Delta H = 0$, $\Delta S > 0$, so only ΔS has any influence on the equilibrium.

If Fig. 15.1 were to be reversed, all that has been said about products would now apply to reactants. Thus we have dealt with two possibilities, $\Delta H = 0$ and $\Delta S > 0$, as well as $\Delta H = 0$ and $\Delta S < 0$.

Endothermic equilibrium

The energy state manifold for an endothermic reaction ($\Delta E_0 > 0$) is shown in Fig. 15.2. Once again we have chosen a 2:1 density of states advantage for the product, B, but, this time, B suffers from an energetic disadvantage as there is an unfavourable energy hill to climb from reactants to products.

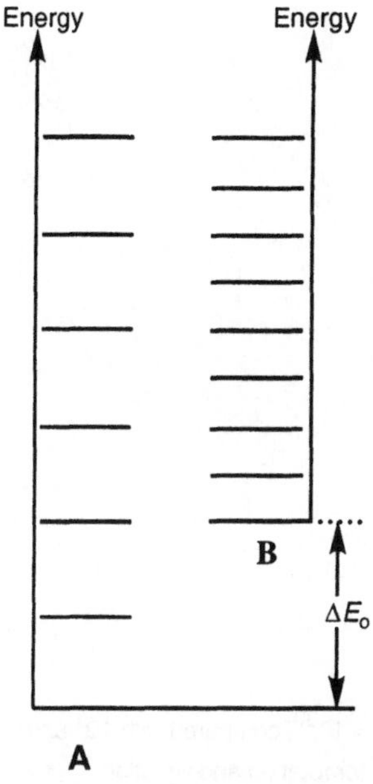

Fig. 15.2 Energy manifolds for an endothermic equilibrium. The 2:1 energy spacing dictates that the equilibrium constant $K \rightarrow 2$ **in the high temperature limit.**

At very low temperatures, the energy is spread only over the low-lying reactant states. At low temperatures, therefore, the dominant feature is the first term in eqn 15.9, $-E_0/RT$. The equilibrium will favour the side of the reaction that is reached with an evolution of heat (energy). Thus the low temperature equilibrium favours the **exothermic direction**. The Boltzmann factor predicts that low-lying quantum states will be the ones that are more populated at **low** temperatures.

As the temperature increases, however, the first term in eqn 15.9 exerts increasingly less influence. As the temperature increases, the energy is spread over increasingly higher energy states, and there are more of these on the product side of the equilibrium. The second term in eqn 15.9, the partition function ratio/entropy term, takes over. Consequently, at high temperatures, side B will dominate. Thus, providing that the **endothermic direction** gives a product with a higher density of states than the reactant, a high temperature will favour the **endothermic direction.** The increased density of states (increased entropy) predicts that higher lying quantum states will be the more populated at **high** temperatures and that is the direction in which the equilibrium shifts.

If Fig. 15.2 were to be inverted, all that has been said about products would, once again, apply to reactants. Increasing temperature would still favour the endothermic direction, which this time is the reverse reaction. Thus, once again, we have covered two possibilities, $\Delta H > 0$ and $\Delta S > 0$, as well as $\Delta H < 0$ and $\Delta S < 0$. The complementary cases, $\Delta H > 0$ and $\Delta S < 0$, and $\Delta H < 0$ and $\Delta S > 0$ merit no special mention since, in both cases, measurable equilibrium cannot be established. For an endothermic reaction with a decrease in entropy, equilibrium will favour reactants at all temperatures, and for an exothermic reaction with an increase in entropy, equilibrium will favour products at all temperatures.

Table 15.1 The effect of increasing temperature on the position of equilibrium for different magnitudes of the enthalpy and entropy changes

	$q_B/q_A > 1$	$q_B/q_A < 1$
$\Delta E_0 > 0$	moves right	stays left
$\Delta E_0 = 0$	stays right	stays left
$\Delta E_0 < 0$	stays right	moves left

Exothermic equilibrium

As we have just noticed, discussion of endothermic equilibrium, also encompasses (by reversal) a discussion of exothermic equilibrium. There is no need, therefore, for a further diagram akin to Fig. 15.2, since all permutations have been covered.

Table 15.1 summarises the results by listing the effect of increasing temperature in various equilibrium situations. It is the value of the quotient of partition functions, q_B/q_A that determines the high temperature tendency of the equilibrium constant; this tendency may be added to or diminished at low temperatures, depending on the sign of ΔE_0. It is the value of this quotient, which represents the extent to which molecules can spread themselves over the available energy states of both A and B, that can be linked to the entropy change for the reaction.

It is always the entropic advantage ($q_B/q_A > 1$) that will dominate at high temperatures, and this will mitigate any energetic disadvantage ($\Delta E_0 < 0$) that may be dominant at low temperatures.

Stoichiometry

Let us consider, finally, the case of an equilibrium reaction with a stoichiometry that leads to an increase in the number of gaseous species of product present compared to reactants. The production of a new gaseous species ensures that there is going to be an increase in entropy, simply because a new mole of gas carries with it such a vast increase in the number of accessible states for the system to enjoy. If we consider the reaction

$$AB(g) \rightleftharpoons A(g) + B(g) \tag{15.14}$$

then we can confidently predict that the quotient of products of partition functions

$$\frac{\prod_{products} \left(q^{\ominus}_{products} / L \right)^{\nu_{pr}}}{\prod_{reactants} \left(q^{\ominus}_{reactants} / L \right)^{\nu_{re}}}$$

$q_{trs} \approx 10^{30}$, compared with 10^1 and 10^0 for rotation and vibration.

will be very much greater than unity. Our confidence arises from the realisation that the creation of a new translational manifold, packed tight with accessible energy states, produces an enormous increase in the partition function for the product side of the equilibrium. Such a reaction will proceed with a significant increase in entropy and will be favoured at high temperatures.

It is the fact that such dissociation reactions are, almost invariably, *endothermic* reactions, requiring an input of energy, which must have led the father of equilibrium, Henri Le Chatelier, to enunciate his well-known principle. Application of this principle to the endothermic dissociation in eqn 15.15, undoubtedly leads to the correctly predicted outcome, but without striking at the root cause of such success. The system responds to an increase in temperature not because bond cleavage in AB soaks up excess thermal energy, but because there is enough thermal energy about to populate the more numerous states on the product side of the equation. At high temperatures, the value of the enthalpy change, and even its sign, becomes an irrelevance. It is always the density of states and entropy that win hands down at high temperatures!

Le Chatelier's principle:
A system at equilibrium responds to any change in conditions by trying to minimise the effect of that such a change.

15.6 Calculating equilibrium constants

We now have all the information needed to tackle the calculation of equilibrium constants for real reactions. We shall consider five separate equilibria, some sharing strong elements of similarity, but each illustrating how to take appropriate short-cuts or use different types of input data.

The five equilibria are:

1. Dissociation of the gaseous dimer of sodium at 1000 K

$$Na_2(g) \rightleftharpoons 2Na(g)$$

2. Isotopic exchange in oxygen at 300 K

$$^{16}O_2(g) + {}^{18}O_2(g) \rightleftharpoons 2\,{}^{16}O^{18}O(g)$$

3. Dissociation of iodine into atoms at 1274 K

$$I_2(g) \rightleftharpoons 2I(g)$$

4. Isotopic exchange between hydrogen chloride and deuterium at 298 K and 1000 K

$$HCl(g) + D_2(g) \rightleftharpoons HD(g) + DCl(g)$$

5. The water gas reaction at 1865 K

$$H_2(g) + CO_2(g) \rightleftharpoons H_2O(g) + CO(g)$$

The data needed to tackle each equilibrium can be obtained from handbooks and from compilations in appendices of good general physical chemistry texts.

In demonstrating how these calculations can be performed, we will run into a number of short-cuts, justifiable approximations, and useful numbers, which, once worked out, will prove to have value again and again in similar calculations.

Data Box

electronic	Na_2	$g_0 = 1$
	Na	$g_0 = 2$
rotation	Na_2	$B/\text{cm}^{-1} = 0.1547$
	Na_2	$\sigma = 2$
vibration	Na_2	$\tilde{\nu}/\text{cm}^{-1} = 159.2$
dissociation	Na_2	$D^{\ominus}_m/\text{kJ mol}^{-1} = 70.4$

$\frac{hc}{k} = 1.439$ cm K

$\sigma_{rot} = 2$ for a homonuclear diatomic

At 1000 K
$V^{\ominus}_m/\text{m}^3\text{mol}^{-1} = 8.206 \times 10^{-2}$

Only the value of q_{trs} is affected by the choice of standard state, so $q^{\ominus}_{trs}$ is written with a standard state symbol.

$mk = MR/L^2$

Short-cut to the value of q_{trs}:
$1.879\times10^{26}(M_r\times T/K)^{3/2}\times V^{\ominus}_m/\text{m}^3\text{mol}^{-1}$
For a dimer,

$q_{trs}(\text{dimer}) = 2^{3/2} \times q_{trs}(\text{monomer})$

$D^{\ominus}_m$ is the dissociation energy from the zero-point level of Na_2 to the SSA (stationary separated atoms) state. It differs from D_0, the dissociation energy from the bottom of the potential well by

$$D_0 = D^{\ominus}_m + \tfrac{1}{2}h\nu_0$$

Problem 1 – The sodium dimer

Determine $K^{\ominus}_p$ for the dissociation reaction at 1000 K

$$Na_2(g) \rightleftharpoons 2Na(g)$$

Data for this problem are given alongside in the data box.

Solution

We start by converting the spectral constants to characteristic temperatures to determine the internal partition functions.

$$\theta_r(Na_2) = \frac{hcB}{k} = 1.439 \times 0.1547 = 0.2226 \text{ K}$$

$$\therefore q_{rot} = \frac{1000}{2 \times 0.2226} = 2246$$

$$\theta_{vib}(Na_2) = \frac{hc\tilde{\nu}}{k} = 1.439 \times 159.2 = 229.1 \text{ K}$$

$$\therefore q_{vib} = \frac{1}{\left(1 - e^{-229.1/1000}\right)} = 4.884$$

$$q_{el}(Na_2) = 1$$

$$q_{el}(Na) = 2$$

Next, some other required data

$$\text{at 1000 K} \quad V^{\ominus}_m/\text{m}^3\text{mol}^{-1} = \frac{RT}{p^{\ominus}} = 8.206 \times 10^{-2}$$

$$q^{\ominus}_{trs}(Na) = \left[\frac{2\pi \times 0.02299 \times 8.314 \times 1000}{\left(6.626 \times 10^{-34} \times 6.022 \times 10^{23}\right)^2}\right]^{\frac{3}{2}} \times 8.206 \times 10^{-2}$$

$$\therefore q^{\ominus}_{trs}(Na) = 5.376 \times 10^{31}$$

$$q^{\ominus}_{trs}(Na_2) = 2^{3/2} \times q^{\ominus}_{trs}(Na) = 2.828 \times 5.376 \times 10^{31}$$

$$\therefore q^{\ominus}_{trs}(Na_2) = 1.521 \times 10^{32}$$

$$\frac{D^{\ominus}_m}{RT} = \frac{E_0}{RT} = \frac{70.4 \times 10^3}{8.314 \times 1000} = 8.468$$

$$\therefore e^{-E_0/RT} = 2.102 \times 10^{-4}$$

$$K^{\ominus}_p = \frac{\left[q^{\ominus}_{trs}(Na)\, q_{el}(Na)\right]^2}{\left[q^{\ominus}_{trs} q_{rot} q_{vib}(Na_2) L\right]} e^{-E_0/RT}$$

$$\therefore K^{\ominus}_p = 2.419$$

Problem 2 – Isotopic exchange in oxygen

Determine $K_p^\ominus$ for the isotopic exchange reaction at 1000 K

$$^{16}O_2(g) + {}^{18}O_2(g) \rightleftharpoons 2\,{}^{16}O^{18}O(g)$$

Data for this problem are given alongside in the data box.

Data Box

vibration	ν_0/Hz
$^{16}O_2$	4.7397×10^{13}
$^{18}O_2$	4.4669×10^{13}
$^{16}O^{18}O$	4.6043×10^{13}

Solution

We start by the writing the factors of the partition function and thence factorise the equilibrium constant $K_p^\ominus$.

$$q = q_{trs}^\ominus \cdot q_{rot} \cdot q_{vib} \cdot q_{el} \qquad\qquad K_p^\ominus = K_{trs}^\ominus \cdot K_{rot} \cdot K_{vib} \cdot K_{el}$$

Reminder:
Only the value of q_{trs} is affected by the choice of standard state, so $q_{trs}^\ominus$ is written with a standard state symbol.

Because there are equal numbers of molecules on each side of the equilibrium, all of the species-independent quantities cancel in factors of $K_p^\ominus$. Thus

$$K_{trs}^\ominus = \left[\frac{m^2(^{16}O^{18}O)}{m(^{16}O_2)m(^{16}O_2)}\right]^{\frac{3}{2}} = \left[\frac{34^2}{36 \times 32}\right]^{\frac{3}{2}}$$

$$\therefore K_{trs}^\ominus = 1.0052$$

Assume bond lengths are unaffected by isotopic substitution, so that values of r in $I = \mu r^2$ all cancel.

$\sigma_{rot} = 2$ for homonuclear, and 1 for heteronuclear, diatomic molecules

$$K_{rot} = \left[\frac{\mu^2(^{16}O^{18}O)\,\sigma(^{16}O_2)\,\sigma(^{18}O_2)}{\mu(^{16}O_2)\,\mu(^{16}O_2)\,\sigma^2(^{16}O^{18}O)}\right] = \left[\frac{(8.4706)^2 \times 2 \times 2}{8 \times 9 \times 1^2}\right]$$

$$\therefore K_{rot} = 3.9862$$

$$\theta_{vib} = \frac{h\nu}{k} = 4.799 \times 10^{-11} \times \nu \qquad \therefore q_{vib} = \frac{1}{\left(1 - e^{-\theta_{vib}/300}\right)}$$

$\frac{h}{k} = 4.7992 \times 10^{-11}$ K s

Isotope	θ_{vib}/K
$^{16}O_2$	2272
$^{18}O_2$	2142
$^{16}O^{18}O$	2208

$$\therefore q_{vib} = \frac{1}{\left(1 - e^{-\theta_{vib}/1000}\right)}$$

$$\therefore K_{vib} = 0.9999$$

First approximation

If all masses, bond lengths, and force constants were identical (while still maintaining a distinction between ^{16}O and ^{18}O), then the equilibrium constant would simply reflect the number of different ways of joining two separate isotopes. There are two ways of choosing an atom from one molecule of $^{16}O_2$ and two of choosing an atom from the $^{18}O_2$ molecule, making four ways in all.

So, all else being equal, $$K_p^\ominus(\text{1st approximation}) \approx 4$$

Second approximation

From cancellation of isotopic masses $K^{\ominus}_{trs} \approx 1$

From the symmetry numbers of the homonuclear species $K_{rot} \approx 4$

From $\theta_{vib}/T \gg 1$ at 300 K $K_{vib} = 1$

From $\Delta\varepsilon_0/kT \ll 1$ at 300 K $e^{-\Delta\varepsilon_0/kT} \approx 1$

So $K^{\ominus}_p(\text{2nd approximation}) \approx 4$

Full calculation

Calculate $\Delta\varepsilon_0/k$ and use this with the values determined earlier.

$\Delta\varepsilon_0/k$ can be calculated directly

$$\frac{\Delta\varepsilon_0}{k} = \frac{1}{2}\frac{h}{k}[2\nu_0(^{16}O^{18}O) - \nu_0(^{16}O_2) - \nu_0(^{16}O^{18}O_2)]$$

$$= \frac{1}{2} \times 4.799 \times 10^2 (2 \times 4.6042 - 4.7379 - 4.4669) = 0.8643 \text{ K}$$

By subsuming the factor of 10^{13} from the frequency directly into the expression for h/k, the more tractable value of 4.799×10^2 is obtained for the constant.

or, more simply, from the characteristic vibrational temperatures

$$\frac{\Delta\varepsilon_0}{k} = \frac{1}{2}[2\theta_{vib}(^{16}O^{18}O) - \theta_{vib}(^{16}O_2) - \theta_{vib}(^{16}O^{18}O_2)]$$

$$= \frac{1}{2} \times (2 \times 2208 - 2272 - 2142) = 0.8643 \text{ K}$$

exactly, but not unexpectedly, as for the direct calculation. This leads to

$$e^{-\Delta\varepsilon_0/kT} = 0.9971$$

$$K^{\ominus}_p = 1.0052 \times 3.9862 \times 0.9999 \times 0.9971$$

So, at 300 K $K^{\ominus}_p(\text{full calculation}) = 3.9954$

Comment

It scarcely seems worth carrying out the full calculation. To within five parts in four thousand, the intuitive and the approximate numerical results point to the same value. But such happy coincidences do not happen with lighter isotopes, as we shall see in Problem 4, where the major mass discrepancy between hydrogen and deuterium does not lend itself to simplistic cancellation.

It is satisfying, nonetheless, to find that straightforward chemical intuition can work so well, and to be able to identify the source of the value 4.

Problem 3 – Dissociation of iodine vapour

Determine $K_p^{\ominus}$ for the dissociation reaction at 1274 K

$$I_2(g) \rightleftharpoons 2I(g)$$

Data for this problem are given alongside in the data box.

Data Box

electronic	I_2	$g_0 = 1$
	I	$g_0 = 4$
rotation	I_2	$\theta_r = 0.0537$ K
vibration	I_2	$\theta_v = 306.6$ K
dissociation	I_2	$\Delta\varepsilon_0/k = 17\,910$ K
molar volume at 1274 K		$V_m^{\ominus}/\text{m}^3\text{mol}^{-1} = 0.1045$

Solution

At first sight, this resembles closely the question posed in Problem 1. However, the data here are more straightforward, and the relatively low value of the characteristic vibrational temperature as well as the high value of the dissociation energy of the iodine molecule provide an insight into the influence of different factors on the final value of the equilibrium constant.

$$q_{trs}^{\ominus}(I) = 1.879\times10^{26}(126.9\times1274)^{\frac{3}{2}}\times 0.1045$$

$$\therefore q_{trs}^{\ominus}(I) = 1.2764\times10^{33}$$

$$q_{trs}^{\ominus}(I_2) = 2^{3/2}\times q_{trs}^{\ominus}(I) = 2.828\times1.2764\times10^{33}$$

$$\therefore q_{trs}^{\ominus}(I_2) = 3.6102\times10^{33}$$

$\theta_{rot}(I_2) = 0.0537$ K

$$\therefore q_{rot} = \frac{1274}{2\times0.0537} = 1.1862\times10^{4}$$

$\theta_{vib}(I_2) = 306.6$ K

$$\therefore q_{vib} = \frac{1}{\left(1-e^{-306.6/1274}\right)} = 4.6753$$

$$q_{el}(I_2) = 1$$

$$q_{el}(I) = 4$$

$\Delta\varepsilon_0/k = 17910$ K

$$e^{-(17910/1274)} = 7.8461\times10^{-7}$$

$$K_p^{\ominus} = \frac{\left[q_{trs}^{\ominus}(I)q_{el}(I)\right]^2}{\left[q_{trs}^{\ominus}q_{rot}q_{vib}(I_2)L\right]}\,e^{-\Delta\varepsilon_0/kT} \qquad \therefore K_p^{\ominus} = 0.1696$$

Comment

The experimental value of the equilibrium constant at this temperature is 0.167, so there is excellent agreement. Note how the very large dissociation energy acts to reduce the equilibrium constant by many orders of magnitude.

Data Box

Compound	
D_2	$\mu = 1.0000$ $\tilde{\nu}/\text{cm}^{-1} = 3118.4$
HD	$\mu = 0.6667$
HCl	$\mu = 0.9726$ $\tilde{\nu}/\text{cm}^{-1} = 2989.74$
DCl	$\mu = 1.8933$

Problem 4 – Isotopic exchange between HCl and D_2

Determine $K_p^{\ominus}$ for the isotopic exchange reaction at 298 K and 1000 K

$$HCl(g) + D_2(g) \rightleftharpoons HD(g) + DCl(g)$$

Data for this problem are given alongside in the data box.

Solution

As in Problem 2, we start by the writing the factors of the partition function and thence factorise the equilibrium constant $K_p^{\ominus}$.

$$q = q_{trs}^{\ominus} \cdot q_{rot} \cdot q_{vib} \cdot q_{el} \qquad K_p^{\ominus} = K_{trs}^{\ominus} \cdot K_{rot} \cdot K_{vib} \cdot K_{el}$$

Because there are equal numbers of molecules on each side of the equilibrium, all of the species-independent quantities cancel in factors of $K_p^{\ominus}$. Thus

$$K_{trs}^{\ominus} = \left[\frac{m(HD)m(DCl)}{m(D_2)m(HCl)}\right]^{\frac{3}{2}} = \left[\frac{3.022 \times 37.46}{4.028 \times 36.46}\right]^{\frac{3}{2}}$$

$$\therefore K_{trs}^{\ominus} = 0.6768$$

Assume bond lengths are unaffected by isotopic substitution, so that values of r in $I = \mu r^2$ all cancel.

$$K_{rot} = \left[\frac{\mu(HD)\mu(DCl)\,\sigma(D_2)\,\sigma(HCl)}{\mu(D_2)\,\mu(HCl)\,\sigma(HD)\sigma(DCl)}\right] = \left[\frac{0.6717 \times 1.906 \times 2 \times 1}{1.007 \times 0.9799 \times 1 \times 1}\right]$$

$$\therefore K_{rot} = 2.595$$

$$\theta_{vib} = \frac{hc\tilde{\nu}}{k} = 1.439 \times \tilde{\nu} \qquad \therefore \theta_{vib}(D_2) = 4488\ K$$

For isotopically related compounds, we assume that force constants are the same, so that all that alters is the reduced mass, μ.

Using $\theta_{vib} \propto (\tilde{\nu})^{-\frac{1}{2}}$, we can find θ_{vib} after isotopic substitution, e.g.

$$\theta_{vib}(HD) = \sqrt{\frac{\mu(H_2)}{\mu(HD)}} \times \theta_{vib}(H_2)$$

$$\therefore q_{vib} = \frac{1}{\left(1 - e^{-\theta_{vib}/T}\right)}$$

	θ_{vib}/K	q_{vib} at 298.15 K	q_{vib} at 1000 K
D_2	4488	1.000	1.011
HCl	4302	1.000	1.014
HD	5492	1.000	1.004
DCl	3083	1.000	1.048

$\therefore$ at 298.15 K, $K_{vib} = 1.000$
and at 1000 K, $K_{vib} = 1.026$

Note that temperature effects in $K_{trs}^{\ominus}$ and $V_m^{\ominus}$ cancel between products and reactants, as do the temperature effects in K_{rot}.
Unlike the case of isotope exchange in oxygen (Problem 2), there are no simple first and second approximations that we can use for this reaction, so we have to go straight to the full calculation.

Full calculation

Calculate $\Delta\varepsilon_0/k$ (= $\Delta E_0/R$) and use this together with the other relevant values determined earlier.

$\Delta E_0/R$ can be calculated from differences between the product and reactant characteristic vibrational temperatures, θ_{vib}, modified by stoichiometric coefficients.

$$\frac{\Delta E_0}{R} = \frac{1}{2}\left[\theta_{vib}(HD) + \theta_{vib}(DCl) - \theta_{vib}(D_2) - \theta_{vib}(HCl)\right]$$

$$= \frac{1}{2}\times(5492 + 3083 - 4488 - 4302) = 0.8643\ \text{K}$$

This leads to

At 298.15 K, $e^{-\Delta E_0/RT}$	$= 1.434$
and at 1000 K, $e^{-\Delta E_0/RT}$	$= 1.113$

$$K_p^{\ominus} = 0.6768 \times 2.595 \times 1.000 \times 1.434 \text{ at } 298.15\ \text{K}$$

$$K_p^{\ominus} = 0.6768 \times 2.595 \times 1.026 \times 1.113 \text{ at } 1000\ \text{K}$$

So

$K_p^{\ominus}$ (298.15 K)	=	2.519
and $K_p^{\ominus}$ (1000 K)	=	2.006

Comment

It is worth inspecting each of the strands that go to make up $K_p^{\ominus}$

- $K_{trs}^{\ominus}$ is temperature independent and much less than unity. The main feature here is the 4:3 mass ratio of D_2:HD.
- K_{rot} is greater than unity simply on account of the ratio of reduced masses, and is then made doubly so because of the symmetry factor (2) of D_2 in the numerator because it lies on the reactant side.
- K_{vib} has little influence on the equilibrium, none at all at 298.15 K and less than 3% at 1000 K.
- ΔU_0 is so small that it points to a virtually thermoneutral reaction. So any effect of temperature on the equilibrium constant will not be great. In fact, the balance of the zero-point energies is (just) exothermic, and the (small) drop in $K_p^{\ominus}$ between 298.15 K and 1000 K is entirely in keeping with the sign and magnitude of ΔU_0.

  Small though ΔU_0 is, its effect is magnified by its appearing in an exponential function. Indeed, essentially all of the decrease in $K_p^{\ominus}$ between 298.15 K and 1000 K can be attributed to this exponential term, entirely as might have been predicted.

Data Box

H_2	$\Delta H_f^\ominus$/kJ mol^{-1} = 0
	$10^{47} I$/kg m^2 = 0.460
	$\tilde{\nu}$/cm^{-1} = 34400
CO_2	$\Delta H_f^\ominus$/kJ mol^{-1} = −393.2
	$10^{47} I$/kg m^2 = 71.4
	$\tilde{\nu}$/cm^{-1} = 2350
	1320
	(2) 668
CO	$\Delta H_f^\ominus$/kJ mol^{-1} = −238.9
	$10^{47} I$/kg m^2 = 14.51
	$\tilde{\nu}$/cm^{-1} = 2168
H_2O	$\Delta H_f^\ominus$/kJ mol^{-1} = −113.8
	I_a $10^{47} I$/kg m^2 = 1.023
	I_b $10^{47} I$/kg m^2 = 1.926
	I_c $10^{47} I$/kg m^2 = 2.956
	$\tilde{\nu}$/cm^{-1} = 3756
	3652
	1595

For hydrogen, carbon dioxide, and steam, the rotational symmetry number $\sigma = 2$; for CO, $\sigma = 1$.

	θ_{vib}/K	q_{vib} at 1565 K	Πq_{vib} at 1565 K
H_2	6332	1.018	1.018
CO_2	3382	1.130	7.629
	1899	1.422	
	961.3	2.179	
	961.3	2.179	
CO	3120	1.158	1.158
H_2O	5405	1.033	1.391
	5255	1.036	
	2295	1.300	

Problem 5 – The water gas reaction

Determine $K_p^\ominus$ for the reaction at 1565 K

$$H_2(g) + CO_2(g) \rightleftharpoons H_2O(g) + CO(g)$$

Data for this problem are given alongside in the data box. $\Delta H_f^\ominus$ is the standard enthalpy of formation of the gaseous species at 0 K.

Solution

This is a reaction with equal numbers of molecules of reactant and product, so, as in Problems 2 and 4, we start by the writing the factors of the partition function and thence factorise the equilibrium constant $K_p^\ominus$.

$$q = q_{trs}^\ominus \cdot q_{rot} \cdot q_{vib} \cdot q_{el} \qquad \boxed{K_p^\ominus = K_{trs}^\ominus \cdot K_{rot} \cdot K_{vib} \cdot K_{el}}$$

Because of equal numbers of molecules on each side of the equilibrium, all of the species-independent quantities cancel in factors of $K_p^\ominus$. Thus

$$K_{trs}^\ominus = \left[\frac{m(H_2O)m(CO)}{m(H_2)m(CO_2)}\right]^{\frac{3}{2}} = \left[\frac{18.02 \times 28.01}{2.016 \times 44.01}\right]^{\frac{3}{2}}$$

$$\boxed{\therefore K_{trs}^\ominus = 13.57}$$

$$K_{rot} = \left[\frac{(\pi I_a I_b I_c)^{\frac{1}{2}} I_{CO}}{I_{H_2} I_{CO_2}}\left(\frac{8\pi^2 kT}{h^2}\right)^{\frac{1}{2}}\right] \times \left[\frac{\sigma_{H_2}\sigma_{CO_2}}{\sigma_{H_2O}\sigma_{CO}}\right]$$

$$= \left[\frac{2(I_a I_b I_c)^{\frac{1}{2}} I_{CO}}{I_{H_2} I_{CO_2}}\right]\left(\frac{8\pi^3 kT}{h^2}\right)^{\frac{1}{2}}$$

$$\boxed{\therefore K_{rot} = 23.54}$$

$$\theta_{vib} = \frac{hc\tilde{\nu}}{k} = 1.439 \times \tilde{\nu} \qquad \boxed{\therefore \theta_{vib}(H_2) = 5968\ K}$$

$$\therefore q_{vib} = \frac{1}{\left(1 - e^{-\theta_{vib}/T}\right)}$$

For a molecule with several modes, $q_{vib} = \prod_i q_{vib}(i)$

$$\therefore K_{vib} = \frac{1.158 \times 1.391}{1.018 \times 7.629}$$

$$\boxed{\therefore \text{at 1565 K, } K_{vib} = 0.2074}$$

Note that temperature effects in $K_{trs}^\ominus$ and $V_m^\ominus$ cancel between products and reactants, but that the temperature effects in K_{rot} do not, because H_2O has one more rotational mode than do the other, linear, molecules.

Full calculation

Calculate $\Delta\varepsilon_0/k$ (= $\Delta E_0/R$) and use this together with the other relevant values determined earlier.

For polyatomic species which have several vibrational modes, $\Delta E_0/R$ cannot be calculated from zero-point energy differences as has been done for diatomic species in earlier problems. Instead, we adopt a direct approach using values of the standard enthalpies of formation at 0 K, as given in the data box.

$$\Delta E_0 = \Delta H_f^{\ominus}(H_2O) + \Delta H_f^{\ominus}(CO) - \Delta H_f^{\ominus}(H_2) - \Delta H_f^{\ominus}(CO_2)$$

$$= -238.9 - 113.8 + 0 + 393.2 = 40.5 \text{ kJ mol}^{-1}$$

This leads to

$$\text{at 1565 K, } e^{-\Delta E_0/RT} = 4.448 \times 10^{-2}$$

So

$$K_p^{\ominus} = 13.57 \times 23.54 \times 0.2074 \times 4.448 \times 10^{-2} \text{ at 1565 K}$$

whence

$$K_p^{\ominus}(1565\text{ K}) = 2.947$$

Comment

The experimental value of the equilibrium constant at 1565 K is 2.8. Experiment (not at all easy to perform at so high a temperature) provides good corroboration for the calculated value. Again, it is worth looking at the reasons why $K_p^{\ominus}$ takes the value it does.

- $K_{trs}^{\ominus}$ is temperature independent and quite large. The dominant feature here is the very low mass of one of the reactants, hydrogen, which appears in the denominator of the equation for $K_{trs}^{\ominus}$.
- K_{rot} is not temperature independent because there is an imbalance in the number of rotational modes across the equilibrium (linear CO_2, H_2, and CO have two rotational modes each, but non-linear H_2O has three). The additional mode is on the reactant side, so K_{rot} will increase with increasing temperature. The value of K_{rot} at 1565 K is quite high because the rotational partition function of carbon monoxide (a product) is much higher than that of hydrogen (a reactant). The other product/reactant pair of gases have much more similar partition functions and tend to cancel each other out.
- K_{vib} has a marked influence on the equilibrium, largely because carbon dioxide, as a reactant, has two modes (the doubly degenerate breathing modes) which are relatively easy to excite and hence are well populated at this high temperature.
- The reaction is distinctly endothermic, so ΔE_0 plays a part in preventing the overall value of the equilibrium constant from becoming very large.

At 1565 K, we have the following (approximate)rotational partition functions :

$q_{rot}(H_2O) = 1020$

$q_{rot}(CO) = 280$

$q_{rot}(H_2) = 9$

$q_{rot}(CO_2) = 1400$

15.7 Conclusions

In this chapter, and specifically in demonstrating that we can determine the position of equilibrium for reactions between simple or even moderately complicated molecules, we have reached the goal that we set out to achieve. The Problems we tackled in Section 15.6 by no means represent a limit to the complexity of the systems to which we are now in a position to apply the powerful methods of statistical thermodynamics. They illustrate a number of lines of approach to general equilibrium situations and the reader is now in a position to explore more complex systems unaided.

However, a modest word of caution is appropriate here. What this book has attempted to do is to lower any barrier that may exist to understanding statistical thermodynamics; that is probably its main intention. Its secondary purpose, or perhaps joint purpose, is to display a little of the fascination and beauty of this subject, to map out the intricate way in which we can build from the simplicity of the concept of quantum energy states to the complexity of understanding some of the detailed interactions and chemical transformations that occur between quite complicated molecules.

The failure of the concept of ideal gas behaviour will become apparent if as simple and important a change as that in the Haber process is tackled using statistical thermodynamics. Gas imperfections, which become important at moderate to high molar densities (such as those at the high pressures used in the production of ammonia by the Haber process), have a marked influence on the position of equilibrium. The reader is encouraged to try some calculations without taking such effects into account, but not to be too disappointed with the results.

Finally, it must be obvious that we have restricted our discussion to the very simplest of cases, to molecules that behave ideally in the gas phase, and to the changes that can occur among such molecules. The real world usually gets rather more complicated than this, yet this book has deliberately shunned such complications. This has been a deliberate choice. There is a wealth of more complete and more rigorous texts available, and the interested reader is encouraged to move rapidly to these as soon as this book fails to supply adequate answers to legitimate questions.

Statistical thermodynamics is not bounded by the limitations we have chosen to place on it here in an attempt to nurture simplicity and foster understanding. Many of the most fascinating aspects of modern statistical chemistry and physics and biology can be tackled by logical extensions to the simple outline presented here. These include

- the problems of gas imperfections,
- the behaviour of condensed matter and the many intricate phase transformations to which it is subject,
- the fascination of a universal description of how all systems approach critical points,
- the significance of the behaviour of mixtures of macromolecules which, on the one hand, form the basis of an important part of the chemical industry and, on the other, promise to provide clues to the complexities of living systems.

All of these bewitching topics, and many more, lie well beyond the remit of an introductory text such as this. The hope, however, is that understanding at this simple level will prove intriguing enough to the reader to kindle, in the future, a need for further enquiry.

Index

OXFORD CHEMISTRY PRIMERS

SERIES EDITORS

This series of short texts provides accessible accounts of a range of essential topics in chemistry and chemical engineering. Written with the needs of the student in mind, the Oxford Chemistry Primers offer just the right level of detail for undergraduate study, and will be invaluable as a source of material commonly presented in lecture courses yet not adequately covered in existing texts. All the basic principles and facts in a particular area are presented in a clear and straightforward style, to produce concise yet comprehensive accounts of topics covered in both core and specialist courses.

Statistical Thermodynamics provides the basic groundwork that will lead third and fourth year undergraduate students of chemistry and chemical engineering from their knowledge of elementary classical thermodynamics to an understanding of the predictable statistical behaviour of assemblies of large numbers of identical molecules, in an ideal gas at constant temperature and volume. It begins by establishing the basis of the Boltzmann distribution law and proceeds, through definition of the molecular partition function, to link the laws of thermodynamics to the statistical behaviour of assemblies of quantum particles. Equations are derived that relate thermodynamic state functions to the molecular partition function and these form a basic toolkit with which to tackle problems from a knowledge only of the relative populations of quantum energy states.

The various contributions to the partition function (translation, rotation, vibration, electronic) are explored and derived. The book ends with a chapter in which all the concepts are brought together in the calculation of equilibrium constants for reactions between ideal gases. A number of fully worked examples are included, making this an invaluable aid to undergraduate chemistry, physics, chemical engineering, and materials science courses. Postgraduate biochemists and molecular biologists will also find this book useful.

Andrew Maczek is Senior Lecturer in Chemistry at the University of Sheffield.

ISBN 978-0-19-855911-5

OXFORD UNIVERSITY PRESS

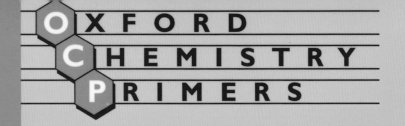